WHEN LIGHT
PIERCED
THE DARKNESS

WHEN LIGHT PIERCED THE DARKNESS

Christian Rescue of Jews in Nazi-Occupied Poland

NECHAMA TEC

New York   Oxford
OXFORD UNIVERSITY PRESS
1986

Oxford University Press

Oxford   New York   Toronto
Delhi   Bombay   Calcutta   Madras   Karachi
Petaling Jaya   Singapore   Hong Kong   Tokyo
Nairobi   Dar es Salaam   Cape Town
Melbourne   Auckland

and associated companies in
Beirut   Berlin   Ibadan   Nicosia

Published by Oxford University Press, Inc.,
200 Madison Avenue, New York, New York 10016

Library of Congress Cataloging in Publication Data

Tec, Nechama.
When light pierced the darkness.

Bibliography: p.
Includes index.
1. World War, 1939–1945—Jews—Rescue—Poland.
2. Holocaust, Jewish (1939–1945)—Poland—Personal
narratives. 3. Christians—Poland—History—20th
century. I. Title.
DS135.P6T29   1985   940.53′15′039240438   85-7261
ISBN 0-19-503643-3

Printing (last digit):   9 8 7 6 5 4 3 2

Printed in the United States of America

To the rescuers,
with gratitude and admiration

p. 39, 37!
32, 36!
13! 209!

Preface

"Yes, speaking," then silence. I introduced myself, explained the study, and told him that since he was a Holocaust survivor who had passed as a Christian, I would like to interview him. I was still explaining when he interrupted me: "My past is my own business, it is too private to be used by anyone . . ." I tried to point to the need, the importance, but there was barely time, for his angry, rough voice cut me short: "You heard me, I will not be a part of your study. My innermost feelings are not for exhibition, not for show . . ." His tone more than his words was a warning. Shaken and embarrassed, I hung up. And yet, I could identify with him.

For three years I lived in Poland under an assumed name, masking my Jewish identity and pretending to be a Catholic girl. I am one of the fortunate few who survived World War II by passing as a Christian. I was sheltered by Christian Poles without whose help my family and I would not have survived.

At the end of the war I resumed my former identity, determined to put the past behind me. I wanted to forget, to forget the person I had so desperately tried to become, to forget that which had forced me to become someone else, and to forget even the Christians who had helped me to stay alive. For years I shied away from wartime memories. Most of my friends, even close friends, knew nothing about my childhood. Nor was my relentless avoidance of the past limited to personal experiences. I stayed away from all readings, all viewing, all discussions of anything even remotely connected to the war. Whenever the subject did come up I kept my silence. Whenever asked a direct question I would answer so evasively that it was obvious no information would be forthcoming. Invariably the subject was dropped and I continued my self-imposed silence.

I succeeded in part, and in part only. Thirty years later my memories began to stir. They called for attention. At first weakly, almost imperceptibly, they began to tempt me to read. Later more forcefully an urge to talk began to assert itself. The need to face and deal with my past became gradually stronger. It was as if I had no choice in the matter. I hardly understood this need, or its power. I only knew that I had to let myself be guided by its compelling force. I decided to revisit my past. I began to write my memoirs, the story of a Jewish girl who was eight years old when the war began and fourteen when it was over.*

As I began to recapture those long-ago experiences the same questions kept recurring. What was it like for other Jews to pass? Which ones had been able to pass, and why? What made some Christians defy all the dangers and risk their lives for Jews, who traditionally were looked upon as "Christ killers," and who, for many still unexplained reasons, were blamed for every conceivable ill? Who were these rescuers? I wanted to know if all those who saved Jews shared certain characteristics. I wanted to know what motivated such people to risk their lives for the persecuted and the haunted.

I felt compelled to know the answers, compelled to share them with as many others as possible. I also realized that my needs to know and to share could not be satisfied by going over my personal experiences alone. My family and I had been rescued by Poles whose main motivation was money; only with time did bonds of affection develop between us. I knew that my ideas about Christian helpers were colored by my experiences. I also knew that one case history, my own, offered only limited insights. I began to search for broader answers. But only a small part of the voluminous literature on the Holocaust deals with rescue, rescuers, and the conditions associated with the self-sacrifice of Christian helpers. Much work would have to be done.†

From the beginning I knew that a trip to Poland would improve my study. My emotions, however, did not necessarily agree with this rational decision. This was 1978. My family and I had left Poland in 1945 under duress, our lives threatened by remnants of the Polish illegal underground. An unsuccessful attempt on my father's life was followed by threats against my sister and myself. Our parents sent us to a different town where we once again had to live under an assumed name.

On a less personal and yet very emotional level, Poland was a Jewish graveyard. Revisiting a graveyard with all its painful memories was not easy. Strangely enough, my apprehension was mingled with attraction to these distant memories, and I vacillated. At that point I knew no one in Poland, so I would be revisiting places not people.

In my most doubting moments I tried to convince myself that this was work I had to do. Yet my doubts persisted. I also suspected that anti-Semitic Poles and the Polish authorities would be hostile toward a Jewish researcher. To guard against such possible complications, a friend who worked in the state department made arrangements with his friend, the then American ambassador in Warsaw, and I was to notify him upon my

---

*See Nechama Tec, *Dry Tears: The Story of a Lost Childhood* (New York: Oxford University Press, 1984).

†Please refer to the Postscript for details on research methods and interviewing techniques.

arrival. This I did, and was directed to an embassy employee, who from then on attended to my needs.

Were my fears justified? Yes and no. When I went to revisit my hometown, I could not find the places I was looking for and those I did find were different from what I had remembered. In this situation I felt threatened and anxious. I had to run away. After two hours I did.

In contrast, when it came to research I was pleasantly surprised. Instead of difficulties, I found cooperation. With some of those I met I established close ties, meeting them socially for dinner and tea. Toward the end of my stay, one of my new friends asked how I intended to transport the many tapes I had accumulated. I had not anticipated any problems with this, although I had been careful not to mention names, so that no one could be compromised. My friend, however, felt that this was not the issue. She was convinced that the authorities at the airport would confiscate the tapes. Independently one of my friends at the Jewish Historical Institute expressed the same fears. They each advised me to use my embassy contact but warned me not to say much over the phone. When I phoned the embassy, as soon as I started saying that I needed a favor, I was quickly interrupted: "I will be in your place tomorrow morning at ten. Does that suit you?" It did, and that finished our conversation.

The embassy employee arrived promptly the next morning. At that point I decided that it would be safest to part with some of my notes as well. Very little explanation was required; my contact knew what to do. I was leaving the next morning, but had one more interview that evening; the American assured me that one or two tapes would create no problems. The rest he would mail by diplomatic post. I must have looked worried, because as he was putting the tapes and notes into a special bag he kept reassuring me that all would arrive safely. It did.

At the airport my luggage was checked and rechecked. Determined to do a good job, the petty official questioned me closely about the name brand of my thirty-year-old wristwatch. I gave him the wrong answer. Suspicious, he became appalled when to his question about money I told him that I never know how much I have and that he was welcome to look. "What, you want me to believe that you don't keep track of your money?" As if offended, nervously he fumbled through my purse, emptying the many compartments and finding $5, $10, $1 bills and more. By now incensed, he mumbled: "Not to know how much money she has, really . . . not to know!" Calmly, I watched him get red in the face. I felt content. I knew that things far more important were escaping his scrutiny.

Westport, Connecticut
July 1985

N.T.

Acknowledgments

"Gratitude is something few of us can afford" said an 87-year-old woman I interviewed in Poland. I identify with the minority she spoke of because I feel deeply indebted to many who in more ways than I can describe helped me start, continue, and complete a project that ultimately became this book.

I am grateful to the Memorial Foundation for Jewish Culture for granting me a Research Fellowship for 1978–79 and 1979–80. I wish also to thank the University of Connecticut Research Foundation for its financial support.

For the Jews and Poles I interviewed for hours, the temporary re-entry into their wartime lives must have been taxing. I am grateful for their willingness to expose themselves to such emotional turmoil. Because I promised all of these research participants anonymity, my wish to thank each of them separately cannot be fulfilled.

I had reached the decision to embark on this project right after I completed a draft of my autobiography. Based on emotional rather than rational considerations this decision moved me into an area I knew little about. Soon I became assaulted by doubts and nagging questions, questions that in addition to answers required outside validation.

Initial determination was followed by uncertainty. Then partial relief came from a highly valued quarter. A former teacher and friend, Robert K. Merton, approved of my project. In a real sense Merton's interest in and support of my loosely formulated ideas gave me the necessary green light.

Important from the start, and throughout my research, was the assistance offered by my former teacher and friend, Herbert H. Hyman. Over the years I have benefited both from Hyman's encouragement and from his constructive criticism.

This study began at the YIVO Institute for Jewish Research, where I was welcomed by a friendly and accommodating staff. I am particularly grateful to my friend, YIVO's senior research historian, Lucjan Dobroszycki, for introducing me to archival research and for sharing with me some of his extensive historical knowledge.

As I searched through different archival collections, YIVO's chief archivist Marek Web was always ready, even eager, to direct me to special, often hard-to-locate materials. In addition to Web, the archivist, Fruma Mohrer, carefully and cheerfully considered all my requests.

Those of us who have worked at YIVO are aware of the broad knowledge and exceptional memory of the chief librarian, Dina Abramowicz. I am indeed indebted to her for saving me many hours of unnecessary searching.

I remember, with a deep sense of gratitude, YIVO's scholarly and gentle historian, the late Isaiah Trunk. A generous man, Trunk urged me to come to him with questions. And when I did, he cautioned me to stay close to the data and to let the data speak. Unfortunately, he is not here to tell me to what extent I have fulfilled his expectations.

A substantial part of this research was conducted in Poland, where assistance came from a variety of sources. When I reached Poland the eminent journalist and historian Władysław Bartoszewski was out of the country. But because I had notified him about my study and my research needs earlier, before Bartoszewski left he instructed his friend, Teresa Preker, also a journalist-historian, to find appropriate interview subjects for my study. Not only did Teresa Preker become a valuable interview contact, but to this day she remains a valuable friend.

Another newly acquired friend, the prominent journalist Marian Turski, helped me in a variety of ways. I am grateful to Turski for his selfless efforts in finding interview subjects, for giving me a special perspective on important historical issues, and for driving me all over Warsaw, a city about which he has an inexhaustible supply of anecdotes.

I was also touched by the particularly warm welcome I received at the Jewish Historical Institute in Warsaw. The chief archivist, Jan Krupka, took a special interest in my research and personally arranged interviews with righteous Christians and Jewish survivors. Krupka also directed me to the appropriate archival collections, while he himself searched for the hard-to-locate cases. I appreciate both the willingness and eagerness with which Zygmunt Hoffman, the historian and vice-director of the Institute, tried to answer my many questions. I also wish to thank Adam Bialecki, the chief librarian, for his valuable reading suggestions. Finally, the Institute's secretary, Edita Mincer, with her hot tea and constant worry that I would catch cold, added a welcome note to my visit.

I have made several trips to Yad Vashem, Martyrs' and Heroes' Memorial Authority in Jerusalem. Each time I came away with a renewed feeling of gratitude and admiration for the chief archivist, Shmuel Krakowski. His interest in my work and the seriousness with which he tackled my many requests will always be remembered. I am also indebted to Mordechai Paldiel, Director of the Department for the Righteous, for his time and valuable information.

Beith Volin in Givataim is a branch of Yad Vashem, housing many important archival collections. Much of my summer in 1979 was spent examining the Beith Volin archives. With gratitude I remember the archivist Sonia Lefkowitz who so readily assisted me in locating appropriate cases.

I am deeply indebted to my friend and colleague, Marilynn Dueker, for her unwavering moral support and ready advice. For years I have been turning to Marilynn for her ability to sort out coding procedures, tables, computers, and data. Her reactions were always generous, always giving.

I wish also to thank my colleague and friend, the historian, Joel Blatt, for his continuous willingness to read and discuss my work. Particularly in moments of doubt, Joel's support and positive attitude meant a great deal to me.

Very important has been the approval and interest of my friend Janusz K. Zawodny. As an expert on Poland, and as someone whose opinion I value highly, Zawodny's positive criticism served as a significant crutch during the more difficult stages of my research.

The involvement, trust, and approval of Susan Rabiner, my editor at Oxford University Press, were truly indispensable to the writing and completion of this book. Susan's care and the time she gave so generously made this book possible. Through our long and frequent working sessions, Susan's contagious enthusiasm turned work into exciting learning experiences, experiences that I am convinced made me work harder. Open, direct, and yet unobtrusive, Susan emphasized again and again that I alone must be responsible for the final decisions. If someone had asked me to describe an ideal editor, I would unhesitatingly point to Susan Rabiner. Both this book and I have gained much from Susan's sensitive guidance and unlimited support.

In this common venture we were frequently joined by Susan's able assistant, Rachel Toor. I have benefited from Rachel's close and careful reading of the manuscript, and I am especially grateful to her for the many interesting questions she raised. I would also like to thank Naomi Schneider, Susan's former assistant, for her initial help and enthusiasm. And I am grateful to Joan Knizeski-Bossert, the preparatory editor, for her careful corrections of the final version of this manuscript.

Only after Oxford University Press published my autobiography did I realize how much hard work goes into an already printed book. I am thinking of staff members who are involved in publicity, sales, foreign rights, movie rights, and much more. Because their work begins after publication, their efforts are not usually acknowledged in the book. To avoid this omission I would like to express my gratitude to the Oxford University Press associates who worked hard after the publication of my autobiography—Laura Brown, Christi Cassidy, Ellen Chodosh, Marjorie Mueller, Peg Munves, Liv Myhre, Helena Schwarz, and Jeff Seroy.

For expert typing and cheerful deciphering of my handwriting I wish to thank Virginia Gough.

All along I have been sustained and nourished by the love and approval of my husband, Leon Tec, and my children, Leora and Roland. During my many exciting discussions with Leora, she urged me not to make this a "dry, scientific" book. I hope that I have not disappointed her.

Roland's reading of the manuscript led to insightful observations and valuable criticism, and I am glad that his overall approval did not interfere with his keen sense for ideas and language.

I have thought of many different titles for this book but was satisfied with none. Instead, I kept postponing the decision. Fortunately, in the last moment I was rescued by my husband who came up with *When Light Pierced the Darkness*. I am delighted that this book carries his suggestion.

While I have so many and so much to thank for, I alone accept the responsibility for all the shortcomings and mistakes.

Contents

WHEN LIGHT
PIERCED
THE DARKNESS

Introduction

There are people the world over who identify with and mourn the fate of Anne Frank. Forbidden from the age of thirteen to venture forth from her hiding place, she nevertheless managed to touch millions. Her courage and her faith in the ultimate redemption of humanity have made her the symbolic Holocaust child. In sharp contrast, very little is known about the Christians who for over two years endangered their lives to protect her. Who were they? What prompted them to undertake this risky venture? What happened to them after the Gestapo discovered Anne Frank's hiding place? Most of these questions were never raised. To this day the answers are not easily found.

The *Encyclopedia Judaica*, for example, devotes more than a page to Anne Frank's story. In this space, however, the only reference to the Christians who tried to save her reads: "From July 9, 1942, until August 4, 1944, the Frank family remained in their hiding place, kept alive by friendly Gentiles. An act of betrayal resulted in their discovery by the German Police."[1] Only a bit more informative is a well-known book about Christians who protected Jews during World War II. The author, without mentioning names, notes that "the two families of noble and self-sacrificing Netherlanders who helped the Franks were sent to concentration camps."[2] A more accurate but limited source, a supplement to Anne Frank's book, explains that when the Gestapo came to Anne Frank's hiding place they arrested two Christians, Mr. Koophuis and Mr. Kraler. After a few weeks Koophuis was released because of poor health, but Kraler spent eight months in a forced labor camp. With no reference to the characteristics or motivations of Anne Frank's protectors, this publication at least gives their names.[3] Only through personal inquiry did I learn that in 1972 Yad Vashem had bestowed upon Anne Frank's protectors the official title of Righteous Christians.[4]

Yad Vashem was established in 1953 in Jerusalem as a memorial to European Jews who perished during World War II, an experience that came to be known as the Holocaust. This memorial also pays tribute to Christians who, during this period, saved Jews. The formal Hebrew title for these saviors is Hasidei Umot Ha-Olam, literally the Righteous Ones of the Nations of the World. In English they are called righteous Christians or righteous Gentiles. Attached to the Yad Vashem institution is a special committee that deals with requests for this distinction. Applications come directly to Yad Vashem or indirectly through a foreign embassy. As a rule, it is next to impossible to receive this title without a request from Jewish beneficiaries[5]; survivors usually petition the committee to honor their protectors.

Depending on the nature and extent of help, special kinds of recognition are bestowed upon Christians who saved Jews. One award is a certificate of honor with a special text in Hebrew and French. Another distinction offers to the recipient a chance to plant a carob tree at the Yad Vashem memorial. Each tree has a plaque with the name of the rescuer. Added to this distinction is a special medallion with the name of the recipient and the inscription: "He who saves one life, it is as if he saves the whole world."[6] These awards are often accompanied by financial aid. Thus, for example, in a poor country like Poland, practically all righteous Christians receive a modest pension. The American Jewish Congress is in charge of all financial arrangements.[7]

To qualify for any one of these distinctions Christian actions had to involve "extending help in saving a life; endangering one's own life; absence of reward, monetary and otherwise; and similar considerations which make the rescuers' deeds stand out above and beyond what can be termed ordinary help."[8] In part ambiguous, these criteria leave no doubt that those who saved Jews because of payment do not fit the definition of righteous Christians. In fact, the Yad Vashem committee ignores all who gained materially by protecting Jews.

If Anne Frank's case is in any way typical, and I suspect it is, this official recognition of righteous Christians is unmatched by any widespread public awareness of, or recognition of, such rescuers.

In the past, the few early publications that dealt with the topic were barely noticed.[9] Only more recently has some interest in the Christians who made Jewish rescue possible become apparent.[10] Thus, for example, the story of the exceptional Raoul Wallenberg has lain dormant for years. Wallenberg, an upper-class Swede, left the safety of his neutral country for conflict-ridden and dangerous Hungary. Early in 1944, in the position of a Swedish diplomat, he went to Budapest and immediately plunged into Jewish rescue, disregarding all warnings and threats. Obsessed with the desire to save, he worked ceaselessly and succeeded in saving thousands of Jews. Wallenberg continued his mission even after the arrival of the Soviet Army. But the Russians, suspicious about his humanitarian motives, arrested him. And even though prisoners who returned from the Soviet Union have testified about meetings with Wallenberg, his captors to this day refuse to acknowledge that he is alive. Raoul Wallenberg has never regained his freedom.[11]

Very different but also late coming to the attention of the public was the case of the German industrialist, Oskar Schindler, a wheeler-dealer, a bon vivant, a man who loved wine and women. Risking his life and ignoring the danger all around him, Schindler saved Jews en masse by employing them in his factory. He managed to escape punishment and survived the war.[12]

This prolonged silence is not surprising. The Holocaust was dominated by extreme suffering and devastation. The compassion and help that were a part of this cruel environment were atypical, easily overshadowed by the enormity of the crimes. It is only natural and expected that those who studied these tragic events focused first on the typical experience rather than the rare exception. Only when this tragic story had been told, would chroniclers begin to notice the less visible, the less obvious: namely, the selflessness and compassion that were expressed in the readiness of some few who were willing to die for others. Once noticed, however, the contrast between the cruelty of the time and the ability of some to rise above it to save the innocent and the helpless emphasizes the nobility of their deeds. Clearly, such heroes, no matter how rare, deserve to be remembered and honored. Apart from paying due tribute, a more systematic study of these righteous Christians would fill a significant gap in the knowledge of the Holocaust.

Such study ought to have additional broader implications. Persecutions, discrimination, and prejudice are a part of our everyday life. More often than not those who are victims of such negative forces cannot effectively fight back. Knowing who would stand up for the persecuted and the helpless, knowing what factors are involved in the protection of the poor, the dependent and the downtrodden, creates an opportunity for cultivating such positive forces.

Moreover, we live in a shaky and uncertain world, a world that offers little help in choosing life values. In such a setting, knowledge and awareness about noble and self-sacrificing behaviors may help restore some shattered illusions. Indeed, mere awareness that in the midst of ultimate human degradation some people were willing to risk their lives for others denies the inevitable supremacy of evil. With this denial comes hope.

For all these reasons, then, the story of those who endangered their lives to save Jews during the Holocaust demands and deserves study. An indispensable prerequisite for such a study is a general understanding of how the Nazi policies of destruction of European Jewry were systematically carried out.

In the absence of other evidence, most historians agree that the plan for Jewish annihilation—the "Final Solution"—crystallized after the start of the Russian-German War in June 1941. As the Germans moved east, they began to search for more efficient ways of extermination. Thus the 1941 capture of Russian territories coincided with the mass murder of Jews, most of which was accomplished by the specially trained SS troops, the *Einsatzgruppen*. For Polish Jews who survived this onslaught the last quarter of 1941 signaled the beginning of the Final Solution. In contrast, only in 1943 did the Nazis decide to move against the Danish Jews by ordering their deportations to concentration camps. To satellite Hungary, the implementation of the Final Solution came even later. The Hungarians were not forced to hand over their Jews until the summer of 1944.

As a general rule, in each country mass murders were preceded by a carefully orchestrated sequence of violations of rights. In the first phase, laws were introduced defining who was and who was not a Jew and requiring the identification of all those who were now defined as Jewish. Next came the expropriation of Jewish property and the denial to Jews of gainful employment.[13] The beginning of the end was signaled by the removal of Jews from their homes to specially designated areas, usually sealed-off ghettoes, out of sight of Christian populations. Isolation of the Jews in the ghettoes before moving them to the death camps was a rigidly enforced part of the master plan in virtually all countries under the Nazi direct control. This was followed by the final mass annihilation.[14]

It was not until the orders came to abandon their homes and move to these specially designated ghetto areas that Jews were in immediate need of help.

In a sense, then, Jewish rescue was a humane response to the Nazi measures of destruction. The appearance of righteous Christians signaled an opposition to, an interference with, the German policies of Jewish annihilation.

Eventually, each European country had some Christians who stood up for the persecuted Jews, but because these anti-Jewish measures were introduced in different places at different times, the timing of the appearance of these righteous Christians also varied from country to country. More importantly, too, the efforts required for Jewish rescue could, and did, vary with time and place. These variations, in turn, can be traced to a series of interrelated conditions that functioned as special obstacles and barriers.

The most formidable barrier to Jewish rescue was the degree to which Nazi occupying forces gained control of the governmental machinery. Where the Nazis were in virtual control, they were prepared to do whatever was necessary to annihilate the Jewish populations and would brook no interference from any individual or group in regard to the carrying out of their policies.

Influencing their decision about how much direct control to exert was their attitude toward the occupied country's Christian population, for in the world of Nazi-occupied Europe, policies and controls depended on racial affinities. For example, the Nazis defined all Slavs and those who lived in the Baltic countries as subhuman, as only slightly above the racial value of Jews. In contrast, the highest racial rank was reserved for the Scandinavians, who bore a closer physical resemblance to the Aryan prototype valued by the Nazis. The other European countries fell somewhere between these two extremes. The Nazis, however, were not always consistent in translating these racial principles into action. Moreover, a particular kind of policy and control, in a particular locality, could and did change with time.

Another condition affecting Jewish rescue was the level of anti-Semitism within a given country. No doubt, within an environment with a strong anti-Semitic tradition, denunciations of Jews and their Christian protectors were more common. In addition, in a society hostile to Jews, Jewish rescue was likely to invite disapproval, if not outright censure, from local country-men. Moreover, in areas of pervasive anti-Semitism, even the helpers themselves could be influenced by long taught anti-Jewish images and values. While they were engaged in the humane act of saving Jews, some must have had to cope with their own negative attitudes about the Jews.

Aside from these cultural patterns, the sheer number of Jews within a particular country and the degree to which they were assimilated must also have affected their chances of rescue. It is easier to hide and protect fewer people. Also, the easier it was for Jews to blend in, the less dangerous it was for others to shield them. And for the rescuers, it would have been easier to identify with those with whom they had more in common. In a sense, too, the extent to which Jews focused narrowly on their Jewishness had a bearing on the extent to which they were differentiated from the native population. These factors, however, came together in an almost limitless number of combinations.

Denmark, for example, represents a very special case; conditions for the collective rescue of Jews were favorable in virtually every regard, and the Danes took full advantage of them. First, Danish Jews numbered only 8,000, making up a mere 0.2 percent of Denmark's population. Second, this small group was highly assimilated.[15] Third, the Nazis defined the Danes as a superior "Aryan" race. Partly because of this definition they were left in charge of their own political destiny, retaining the prewar government.[16] Indeed, of all European countries under Nazi domination, Denmark en-joyed the most favored position, becoming Hitler's model protectorate state, which functioned relatively undisturbed until 1943.[17] One effect of Den-mark's local autonomy was that the Jews were left alone.

As long as no Christian Danes were threatened for dealing with Jews, the idea of Jewish rescue was irrelevant; it was precisely the minimal interfer-ence of the Nazis in the internal affairs of Denmark that made the idea of a righteous Christian superfluous.

Only in 1943, when the Nazis insisted on applying the Final Solution to Danish Jews, did the situation change. The Danes refused to obey. The Nazis reacted by taking over the government, and after gaining control they tried to implement their anti-Jewish policies. Only at that point did the idea of righteous Christians become a reality.

Traditionally, the Danes had perceived the Jews as Danes, and now they refused to hand them over to the Nazis, even when ordered to do so. At this point the Danish underground gained in stature and, with the coopera-tion of other Danes, moved the country's Jewish population to Sweden.[18]

Danish opposition to the Nazi measures saved even the 472 Jews who were seized by the Germans before they could be evacuated and sent to the concentration camp Theresienstadt. The Danes insisted on visiting them, and the Nazis went along with their demands. Compared to their treatment of other Europeans, the Germans were lenient toward the Danes, even after rescue attempts started. Reprisals took the form of arrests and usually applied only to the organizers of the Jewish exodus.[19]

A low degree of anti-Semitism facilitated the Danish protection of Jews, as did their strong adherence to democratic principles. No doubt, too, the fact that the Danes saw their Jews as Danes helped them identify with the Jewish plight. For the very same reasons these Christians experienced little approbation from other Danes, nor did they have to cope with personal anti-Jewish images and values.[20]

Holland represents a less clear-cut situation than Denmark. Right after the Germans occupied the Netherlands they installed a civil administration headed by the highly experienced Nazi official, Artur Von Seyss-Inquart. As Reichscommissar he wielded absolute power, which he applied ruthlessly to the implementation of anti-Jewish policies.[21]

In 1941 as the initial anti-Jewish measures were being introduced, the citizens of Amsterdam protested by calling for a general strike that lasted three days. The Germans reacted swiftly with severe reprisals against both Christians and Jews. From both groups many arrests and deportations to concentration camps followed. In this confrontation the Nazis came out as the definite winners. In the climate of resignation that permeated Europe at this time, when even the great powers of the continent seem to have conceded the irresistibility of Nazism as the future world force, such an outcome seemed inevitable. Well-organized Dutch opposition did not resume until 1943.

In 1942, after a series of preliminary anti-Jewish measures, the Nazis began to implement the Final Solution in Holland. This involved concentrating most of the Jews in the Camp Westerbrog, from which they were moved to concentration camps in Poland, usually Auschwitz. Those deportations created the need for Christians who were willing to protect the Jews.

Aware that their plans could be sabotaged by the Dutch, what measures did the Germans take? In general they made it clear that they "considered assistance to the Jews as help to the enemy which was punishable by torture and deportation to concentration camps."[22] Still, it is also true that the Nazis were not always eager to inflict the heaviest penalties on the Dutch who were shielding Jews. They wished to avoid confrontations with the native population. Thus, for example, a farmer was caught hiding Jewish children on his farm three times; only after the third time was he sent to a concentration camp, where he died.[23] On the other hand, some of those who rescued Jews were executed. This happened to the known underground figure Joop Westerweel, who worked tirelessly on behalf of Jews. He was caught, tortured, and murdered.[24]

By 1943 the Dutch underground was well organized. At that point its leaders came primarily from rural areas; many of them were active church members and some admitted to a personal aversion to Jews.[25] Yet, despite mixed attitudes toward the Jews, the Dutch underground equated the aid to Jews with opposition to the occupying forces. Of the Jews who survived by passing as Christians, many benefited from the underground's aid.[26] However, the deportations in Holland started in 1942 whereas the Dutch underground became effective only in 1943. For an entire year, then, the Jews were deprived of the aid of the only force organized to resist the Nazis. Here, clearly, timing acted to the Jews' disadvantage.

As a group the Jews made up 1.6 percent of Holland's population, or 140,000 individuals. They were relatively well integrated into the society. Regarding the degree of Dutch anti-Semitism, there are some differences of opinion. The Dutch were thought by some to harbor the same low level of anti-Semitism as did the Danes,[27] but others with personal experiences as Dutch Jews have disagreed with this assessment.[28] The difference between the two peoples may be related to the fact that conditions, as discussed above, made it easier for the Danes to be righteous when it counted most for the Jews.

Similarly, reports of the number of Jews who benefited from Dutch help also differ. All agree, however, that a substantial proportion of those who tried to survive among Christians were denounced. Raul Hilberg maintains that of 25,000 who lived illegally among the Christians, 7,000 survived.[29] Others have stated that of the 24,000 who hid in the Christian world, 16,000 succeeded in eluding the enemy. Unlike Hilberg's figures, the second set includes those who were married to Christians and those who were exempt from deportation for other reasons.[30] Whatever the reasons, the Dutch did not succeed in saving many Jews. Of the Jewish population in Holland, more than three-quarters perished.[31]

Another illustration of interest might be the very different case of Bulgaria. The wartime behavior of the Bulgarian government has been described as opportunistic. To reclaim territories lost during World War I, Bulgaria joined forces with Germany quite early on and became a satellite state. The same spirit of opportunism is said to have prevailed within the government in its policy toward the Jews. Only later, under pressure at home, and because the Germans began to lose the war, did the government reverse its anti-Jewish policies.[32]

Throughout the war, Bulgaria retained its independence and the Germans refrained from direct intervention in the internal affairs of the country.[33] This, however, did not prevent them from influencing and even pressuring the Bulgarians into anti-Jewish actions. Well before the war, in the 1930s, the Nazi influence was introduced when some Bulgarian students came back from studying in Germany, imbued with the anti-Semitic Nazi ideology. They began as students to advocate anti-Jewish measures. The results were boycotts of Jewish businesses, acts of discrimination at institutions of higher learning, and overall agitation that expressed itself in occasional violence against the Jews. On the whole, however, attempts at stimulating anti-Semitism met with no more than moderate success. The public showed little enthusiasm for such hostile measures.[34]

But by 1941, with Germany at its peak of conquest, the Bulgarian government began to introduce the package of anti-Jewish laws that had to do with the definition, expropriation, and concentration of Jews.[35] The public disapproved of these steps. Thus, for example, the law requiring Jews to wear the yellow Star of David was attacked with special vigor. Under pressure, the government had to back down. And as the Bulgarian authorities continued to move closer to deportations, voices of opposition grew louder. Protest came from several directions: the Bulgarian Orthodox Church, the intelligentsia, the Communists, the underground, and even some segments of the government.[36] The Orthodox Church in particular became very active in its opposition to Jewish persecution. Some priests performed mass conversions, often falsifying the dates, while others performed hasty marriages between Christians and Jews. All these actions shielded the Jews from growing dangers. The priests' zeal, however, did not pass unnoticed. The government reacted by censuring some priests and by removing others from their jobs.[37] Still the church was not intimidated. In September 1942, the Metropolitan Stefan of Sofia delivered a sermon in which he stated that God had already punished the Jews "for having nailed Christ to the Cross," by driving them from place to place, and by depriving them of their own country. It is God, he insisted, and not men, who determines the Jewish fate. Men, therefore, have no right to persecute them.[38]

By 1943 the tide of war seemed to have turned. A German victory was no longer at all certain. The Bulgarian government could not clearly define a path of least resistance. The public, forming a powerful coalition, continued to demand a stop to anti-Jewish actions. The Nazis, on the other hand, continued to pressure Bulgarians into delivering their Jews. The government compromised by offering to the Germans the "foreign" Jews, those who resided in the recently annexed territories: Macedonia and Thrace. As these foreign Jews were being deported, the opposition continued to grow. Once more the head of the Orthodox Church intervened on behalf of the Jews, this time for Jews who resided in the freshly acquired territories. Added to this intervention was the powerful voice of Dimitur Pleshev, the vice-president of the Bulgarian parliament. Pleshev headed a delegation that approached the Ministry of Interior demanding a stop to the deportation of these Jews. When this effort failed, Pleshev introduced in the parliament a resolution of censure accusing the government of complicity in the atrocities that had occurred during the deportations. Pleshev was voted down and subsequently lost his position. None of these protests were able to alter the fate of foreign Jews. The government delivered to the Germans 11,384 Jews from Macedonia and Thrace. Practically all perished in the concentration camps in Poland.

These protests, however, may not have been made in vain. They were followed by a high order, presumably from the Bulgarian king, forbidding all Jewish deportations from Old Bulgaria.[39] Thus, the Jews from Bulgaria proper were spared. Half of the Bulgarian Jews lived in Sofia, the country's capital; at one point they were expelled to the provinces, but under pressure even this order was revoked. Because of continual internal unrest, by 1944 the government changed hands and the new regime nullified all anti-Jewish legislation.

The timing of these anti-Jewish actions, but more significantly the absence of direct Nazi control, facilitated the saving of Bulgaria's Jews. Throughout, the relative independence of the Bulgarian government functioned as a buffer. Furthermore, the limited number of Bulgarian Jews, 50,000 or 1 percent of the population, might have also improved their chances of survival, while their low level of assimilation worked against it.

Historically, the Jews of Bulgaria had enjoyed full political and economic freedom. Culturally and communally, however, they were segregated from the rest of society, with the majority of their children attending Jewish schools.[40] On the whole, the Bulgarian Jews played no significant role in the country's cultural and political life. Jewish achievements were modest, and partly because of this, serious clashes or competitiveness were absent from Christian–Jewish relationships.[41]

Because the Nazis had no direct control over local affairs, Bulgarians who stood up for Jews and Jewish rights were not subjected to severe punishments, and thus were not endangering their lives. Yet, without them many more lives would have been lost. By no means clear-cut, the Bulgarian experience represents a very special and interesting case.

In contrast to the examples thus far described and for many of the reasons mentioned, obstacles and barriers to Jewish rescue were the most formidable in Poland. Poland was both extreme and special.

Quite early, the Nazis designated Poland as the center for Jewish destruction. Whether in so doing they were prompted by the country's concentration of Jews or whether they counted, if not on actual support, at least on the indifference of Poles, or on the absolute subjugation of the Poles, we will never know. For this decision the Germans left no records, no explanation. What they left were records of unprecedented human destruction. The estimates for Jewish survivors in Poland range from 50,000 to 100,000, compared to the prewar population of 3.5 million.[42,*]

As the center of Jewish annihilation, Poland provides the key to an understanding of the Holocaust in general and to the rescuing of Jews by righteous Christians in particular. As a country in which the Holocaust drama was played out in the most gruesome and ruthless ways, Poland can teach us about similar but less extreme cases. Partly for these and partly for personal reasons, this book focuses on Poland.

Convinced as I am that the study of Poland would offer general theoretical insights, I have divided this book into three parts. The first four

*The estimate of 50,000 to 100,000 survivors refers only to Polish Jews who remained in Poland. During the war an estimated 250,000 managed to escape to Russia, Sweden, Switzerland, the United States, and Palestine. Of this number, the majority went to Russia.

chapters describe what it was like to pass and hide among Christians and what it was like to rescue Jews. Chapters 5 and 6 describe two exceptional cases of help—that of the paid helper and that of the anti-Semitic helper. The remaining chapters try to account for righteous aid by examining the characteristics of righteous helpers, their motivations, and the conditions associated with such rescue. In each part I rely on the literature and my own research.[43] Important and interesting, the bulk of the Holocaust literature about righteous Christians consists of scattered case histories and personal accounts. While offering occasional theoretical insights, such publications contain no overall systematic and comprehensive explanations.[44]

What were the overall conditions in Poland? How conducive were they to Jewish rescue?

In 1939, of all the European countries, Poland had the highest concentration of Jews. They made up 10 percent of the country's population. As the largest community of Jews in Europe, Polish Jews were also the least assimilated. They looked, dressed, and behaved differently from Polish Christians. Some of these differences can be traced directly to religious requirements that called for special rituals and dress. Others were accentuated by the urban concentration of Jews. Over 75 percent lived in urban settings, while the same was true for only 25 percent of the Polish population in general.[45] Urban–rural differences were magnified by occupational differences. Of the Jews, only 4 percent were engaged in agricultural pursuits, while 79 percent were employed in manufacturing and commerce. In contrast, for the rest of the population, over 60 percent lived off agriculture and 25 percent were employed in manufacturing and commerce.[46]

Educational experiences also helped to exaggerate the existing distinctions. In prewar Poland more than half the Jewish children attended special Jewish schools.[47] Enrollment in religious school, in turn, discouraged mastery of the Polish language. Thus, in answer to a 1931 census inquiry, the overwhelming majority of Jews mentioned Yiddish as their native tongue (79 percent) and only 12 percent gave Polish as their first language. The rest chose Hebrew.[48] Jews and Poles lived in separate and different worlds,* and their diverse experiences made for easy identification. It has been estimated that more than 80 percent of the Polish Jews were easily recognizable, while less than 10 percent could be considered assimilated.[49]

Apart and different, Jews are first mentioned in Polish history in the tenth century, an era that coincides with both the emergence of Poland as a nation and Poland's acceptance of Christianity.[†] In and of itself, this late acceptance of Christianity contributed to the church's insecurity. The Jews, by rejecting the new religion, seemed to deny its legitimacy. Largely be-

---

*Partly because Jews and Christians in Poland were very different and lived in separate worlds, and partly for convenience, I refer to Polish Jews simply as Jews. For the same reason a Pole is a Christian who lives in Poland. When referring to Poland, I use Pole and Christian interchangeably.

†What follows is a brief and incomplete history of Polish-Jewish relations. A more comprehensive treatment of the subject is beyond the scope of this book.

cause of the church's insecurity, Jewish nonacceptance of the new religion became a special threat and challenge. The church reacted by turning against Jews and by defining them as "Christ killers." In adopting an anti-Semitic posture the church followed the example of other countries. Indeed, soon in Poland, as elsewhere, to the view of the Jew as Christ killer was added the myth of the Jewish ritual killing of Christians. This ritual, they assumed, took place in order to use Christian blood for the making of matzo. As is often the case, ignorant ideas were followed by ignorant actions. As early as 1399, the Jews of Poznan were accused of killing three Christians for the purpose of using their blood for matzo. At the trial, thirteen Jewish elders were sentenced to death by burning. In Cracow, eight years later, a priest announced that Jews had murdered a Christian child. This announcement was followed by a large-scale massacre, plundering, and the burning of Jewish property.[50]

But neither the attitudes toward Jews nor their treatment was always consistent.* The economic importance of Jews was first recognized by the medieval princes who had invited them to settle in Poland. This invitation was motivated by a desire to improve the country's backward agrarian economy. Also, because the church defined usury as sinful, many Jews began to fill an important economic function by becoming Poland's official money lenders. Others were given minor administrative posts at the estates of the princes. Because Jews were barred from ownership of land, most of them had to become merchants, artisans, and professionals. Since these occupations required an urban setting, Jewish presence fulfilled the princes' wish for modernization. Aware of the Jews' economic usefulness, the powerful nobles tried to protect them from growing religious opposition. Their efforts, however, became unsuccessful when to the hostility of the church was added resentment due to economic competition. The rising class of Christian merchants objected to their Jewish competitors, as did the minor nobility. Both aspired to the positions held by the Jews. To the resentment of these groups was also added the resentment of peasants who had direct dealings with the Jewish estate administrators and who blamed them rather than the lords for all their misfortunes.[51]

With time religious and economic anti-Semitism found an ally in political anti-Semitism. And eventually this political anti-Semitism assumed a prominent umbrella-like position, dominating and directing anti-Jewish moves into all other areas.

How did this come about?

Surrounded by Austria and Prussia and by Russia, for centuries Poland had to defend itself against its powerful and aggressive neighbors. Continuous external threats coincided with Poland's internal dissension and divisiveness. Ultimately chronic internal dissension together with equally chronic external threats led to territorial losses. The winners were Poland's

---

*Thus, for example, the fourteenth century Polish King Casimir the Great made Poland a haven for Jews by extending to them far-reaching privileges.

ever-eager neighbors: Austria, Prussia, and Russia.[52] With time these three powers succeeded in dividing and taking over the country. By 1795 Poland had ceased to exist as an independent state; only in 1918, after World War I, did it regain its sovereignty. Characteristic of the newly emergent state were its internal political dissension, severe economic problems, and an overall lack of cohesion.

Of more than 30 million Polish citizens, one-third belonged to diverse minority groups. By far the largest of them were the Ukrainians, amounting to 4.5 million. Next in numbers came the Jews with almost 3.5 million. The Belorussians made up 1 million and the Germans 740,000. Restless, these minorities showed little loyalty to the new state. In fact, except for the Jews, none were satisfied with their new citizenship; each aspired to and identified with a different nationality. Ironically, even though of all the other groups the Jews were most satisfied with and loyal to the new country, they, rather than any of the others, paid a price for the new nation's economic and social problems.

The rebirth of the new state started with a wave of destructive anti-Jewish outbursts. From November 1918 until January 1919, 100 violent incidents occurred against Jews.[53] Those who bothered to justify this violence explained that Jews were communist spies and that these acts served as preventive measures against communism.[54] Disregarding reality, the identification of Jews with communism continued to function as a powerful reinforcer of anti-Semitism.[55]

Because of these excesses the Poles were made to sign the 1919 Versailles Minority Treaty, which guaranteed the rights of all minorities. This meant that the government could not introduce legislation that curtailed the rights of its minorities. Yet, absence of discriminatory laws did not guarantee equality.[56] On the contrary, Poles refused to abide by any anti-discriminatory clauses. They perceived the Versailles Minority Treaty as an insult to their national honor, as an internal interference. Violence against the Jews became a way to defy a perceived affront to their national dignity.[57] The Versailles Minority Treaty failed to fulfill its promise.

Judging from the progressively deteriorating economic conditions, anti-Jewish excesses also failed to improve the country's economy. Nor did they result in the much needed national unity. In 1925 Poland officially registered ninety-two different political parties.[58] Between 1918 and 1925 the country was ruled by fourteen separate governments.[59]

Such internal upheavals were temporarily halted by Marshal Piłsudski's 1926 coup d'état. As head of the movement, Sanacja, Piłsudski established a unique dictatorship. The parliament, including numerous opposition parties, still retained some of its influence. In fact, Piłsudski's government was surrounded by many opposition blocks, each of which readily voiced its demands.

The opposition to the right of the political spectrum consisted mainly of the National Democratic Movement (SN or Endecja),* its head the conser-

*The National Democrats are also referred to as the National Party, which is a translation from the polish *Stronnictwo Narodowe* (or SN).

vative Roman Dmowski. As a dedicated anti-Semite, he believed in the supremacy of the Polish nation and church. The Jews, he felt, were a foreign element, and therefore a threat to the coherence of national institutions.

The opposition to the left of the political spectrum consisted of the Socialists (PPS), the National Workers Party (NPR), and the Populist Party, which represented the poorer peasants.[60] Before seizing power Piłsudski belonged to the Socialist Party, which was ideologically opposed to anti-Semitism. As head of state, Piłsudski broke his ties to the Socialist Party but continued to oppose anti-Jewish measures. While Piłsudski was alive he was able to resist successfully most of the anti-Semitic demands of the National Democrats.[61] And, as a strong and charismatic leader, he succeeded in temporarily reconciling some of the conflicting political interests and demands.

In 1935, with Piłsudski's death, the situation changed drastically. His successors had a hard time holding on to power. Their own camp (Sanacja) was in a state of disarray. They did not have Piłsudski's broad base of support within the population at large. Economic and social conditions in the country deteriorated further. At the same time, the National Democratic movement kept pushing the government toward a stronger anti-Semitic position.

Eventually, in 1937, Piłsudski's successors set up the Camp of National Unity (OZON), which became the chief base of the semidictatorial regime, rallying the nation around the army, nationalism, and Catholicism.[62] They elected to try to solve the country's problems by openly supporting anti-Semitism in social, cultural, economic, and political areas.[63] Their policy led to a proliferation of anti-Jewish measures, which extended to most societal realms.[64]

In the economic sphere these restrictions led to a variety of concrete actions, one of them aimed at taking over Jewish market stalls. Another prevented Jews from holding civil service jobs. Jews were also barred from employment in the state-owned monopolies, such as the liquor and tobacco industries. In addition, the government introduced examinations for artisans especially designed to fail Jews (those who could not pass the examinations were barred from employment)—a measure that posed a serious threat to the livelihood of the Jewish artisans.[65] Also through special pressures, Polish unions succeeded in limiting Jewish employment even in the textile mills owned by Jews.[66]

The National Democrats were also in favor of boycotting Jewish businesses, and Sanacja leaders, without official endorsement, went along with their wishes. Anti-Semitic youths stood in front of Jewish stores preventing customers from entering. Window breaking and beatings accompanied such measures.

At times what had begun as a business boycott would turn into a general pogrom. This was indeed the case with the 1936 Przytyk pogrom, which led not only to property destruction but also to loss of life.[67] Less evident boycotts of Jewish businesses were supported through a proliferation of

propaganda leaflets, placards, and posters. All such printed material condemned Jewish businesses. A typical poster read as follows (see above):

"Fellow countrymen, we are being murdered by Jews and yet we keep supporting
them. We give them money to fight us. This is a shame and a disgrace. Polish
consciousness awake! Let the innocent blood of our comrades stand before you
when you are about to commit a deed calling for God's revenge, when you are
about to carry money to the Jew. You, our countrymen and our Polish mothers,
remember: avoid your enemy the Jew. Support only your own kind!"

At times such calls to action appeared together with vivid pictures. Thus, for example, one of the posters depicts those who buy from Jews as pigs and donkeys (see below). In contrast, those who buy in Christian establishments look like decent citizens who follow the national interests of their country. The caption under the drawing says: "Note who buys from Jews! Which side are you?"[68]

As with employment restrictions, the Jews suffered from these boycotts. A survey of ten small towns that experienced business boycotts documents the tangible losses. In the years 1932–1937 these towns had experienced a 28 percent drop in Jewish businesses.[69]

Discriminatory practices were by no means limited to employment and commerce. Institutions of higher learning also became battlegrounds for anti-Semitism as they restricted or eliminated Jewish participation. Universities that accepted Jews reserved separate benches for them. Prohibited from sitting anywhere else and unwilling to use the assigned spaces, the Jewish students stood in protest during all lectures. They were also physically threatened and abused. During the period 1937–1938 some universities had to suspend classes solely because of the violence committed against the Jews. Between 1925 and 1939 the proportion of Jewish students at Polish universities dropped from 21.5 to 8.2 percent. This decline was particularly marked in later years. In the last four years preceding World War II the population of Jewish students was reduced by 3,000.[70]

Taking their cue from their Nazi neighbors, the Polish government be-

Przypatrz się, kto kupuje u żydów i... po której stronie ty jesteś.

gan to search for a so-called Jewish solution.[71] In 1937 Józef Beck, the foreign minister, speaking in the Polish parliament, insisted that Poland had space only for half a million Jews. This meant that almost 3 million had to leave.[72] The Poles showed great interest in the Nazi Madagascar plan, aimed at removing all Jews to the island off the coast of Africa.[73] The government's desire to be rid of the Jews and the seriousness with which it was pursued was shown in a 1938 Polish mission to Madagascar.[74]

As ideas about Jewish removal were gaining momentum, realistic chances for their fulfillment were becoming more remote. Barriers for Jewish emigration had emerged from a variety of directions. A white paper promulgated by the British severely limited Jewish influx into Palestine, an action taken in part as a gesture of goodwill toward the Arabs.[75] By 1924 the United States' traditional receptiveness to immigrants underwent a drastic change, making it the last year for open immigration. In the late 1930s the United States accepted fewer than 7,000 immigrants from Poland, a figure that included Poles as well as Jews. Largely because of an international economic crisis, many countries, among them South American ones, introduced immigration restrictions.[76]

Polish Jews faced a difficult situation. Impoverished, with no choices, no place to go, unwanted, they continued to stay. Time was working against them. Jews were faced with a progressively deteriorating situation.

Only when in September 1939 Poland was attacked by Germany did the common danger create a feeling of unity between Poles and Jews. However, the Germans made sure that this Polish–Jewish cooperation would not last. Almost immediately after they invaded Poland the Nazis started to loot Jewish homes. This they did with the help of the Polish anti-Semitic mob, who served as informers.

Among the early Nazi successes was the instigation of a number of pogroms in which the Polish underworld and Fascist elements participated. The best known among these was the February 1940 pogrom in Warsaw. It lasted three days.[77] These seemingly spontaneous outbursts of violence were in reality carefully orchestrated by the Nazis. Introduced in different places at different times and with different degrees of ruthlessness, such physical attacks represented but one of many that eventually led to Jewish extermination.

In 1939 when the Germans occupied Poland they had behind them six years of experience with anti-Jewish policies. This experience added to their efficiency. Moreover, the traditionally strong Polish anti-Semitism with its many anti-Jewish policies offered a receptive environment for all other anti-Jewish measures.

No doubt the Nazi view of Poles as only slightly above the Jews on the scale of racial values gave them a free hand to act in any way they saw fit. Thus, for example, whereas in Germany the murder of a part Jew (*Mischling*) would mean killing a part German, this did not apply to Poland.[78] By murdering a part Pole the Nazis were contributing to the elimination of two "undesirable" elements, and thus "purifying" the world's racial composition. Contemptuous attitudes toward Slavs helped remove all potential constraints. In Poland, then, without consideration of human suffering, the Nazis began to introduce anti-Jewish policies that culminated in extermination, the Final Solution.

How did this come about?

In 1939 after Poland's defeat, the Germans fulfilled their treaty obligations to the Soviet Union by allowing the Red Army to advance to what was called the Curzon line. Another part they annexed to Germany, and in the remainder they set up a Nazi government headed by a governor general, Hans Frank. As head of the newly truncated Poland, Frank moved his headquarters to the ancient capital, Cracow. The establishment of a Nazi-controlled Poland was followed by a quick succession of new laws. After the initial outbursts of anti-Jewish violence, special documents were issued to Jews identifying them. Jews were required to wear the Star of David either sewn into their clothing or more frequently as part of a special armband. Travel restrictions were followed by prohibitions of movement within communities. This last measure prohibited visits to the more desirable sections of most communities.

Also in 1939 the Nazis ordered the prewar Jewish self-governing bodies to form special councils or *Judenrat*. At that point the Germans assured the Jews that the councils would mediate between them and the Jewish constituents. In reality they expected these Jewish councils only to follow orders. At first the councils had no idea what was in store for them or for their people. Some obeyed the German directives. Those who refused, paid with their lives.

Faced with an unprecedentedly legitimized anti-Semitism, unable to draw upon past experience, with no power to which to appeal, members of these councils reacted in a variety of ways. In the end, no matter how they behaved, all were destined to die. The lives and deaths of these wartime council members represent a tragic and controversial chapter in the history of the Holocaust.[79]

Pointing to the initial outbursts of anti-Jewish violence, the Nazis presented it as proof that Christians and Jews must be separated. They argued that placing Jews in ghettoes would do away with racial conflicts.

The formation of ghettoes was preceded by the forceful removal of Jews from one locality to another. The purpose of these relocations was to concentrate as many Jews as possible in as small an area as possible. This goal the Nazis achieved through a total disregard for peoples' physical, psychological, and social needs. The first ghetto was formed in Piotrkow in October 1939, only to be abandoned by the Nazis without any explanation. This was followed by the establishment of the Lódź ghetto in 1940, which included 160,000 inmates. The year 1940 saw the creation of the Warsaw ghetto with a population of about half a million. Lódź, Warsaw, and Bialystok, the three largest ghettoes, contained 1 million Jews, about one-third of the Polish Jewry.[80]

By 1941 practically all Polish Jews lived in separate communities. Almost without exception these were located in the most dilapidated parts of urban centers. Isolated from the rest of the population and from each other, the inmates in these ghettoes were forced to work for starvation wages in industries that supported the German war efforts.

Orders to move into ghettoes were issued with total disregard to human suffering, and failure to comply carried the death sentence. At first these ghettoes contained only Jews from surrounding areas, villages, and smaller towns. Soon their numbers were augmented by arrivals from other European countries. Eventually the Lódź ghetto came to include even Gypsies.[81]

One of the Nazi Nuremberg laws specified that whoever had three Jewish grandparents was Jewish. This meant that those Jews who had converted to Christianity were counted as Jews. All these "Catholic Jews" were ordered into the Warsaw ghetto. There they were assigned special quarters and given permission to practice their religion.[82] Not only had these converts rejected Judaism, but in the process some of them had become avid anti-Semites. Understandably, therefore, the rest of the ghetto population saw them as traitors and found their presence and special "privileges" objectionable and humiliating. Thus, for example, one of these "ex-Jews," Szeryński, became notorious as the head of the Jewish ghetto police. The Jewish underground tried to kill him but succeeded only in injuring him. In the end, in 1943, during the final stage of the Warsaw ghetto, Szeryński committed suicide.[83] However, the most visible and most troubling ghetto features were overcrowding, hunger, filth, epidemics, and high death tolls. One ghetto inmate who later perished in a concentration camp notes that the Nazis *began the slow and gradual killing of our people . . . . Public notices on gates and walls constantly announced bans and cruel decrees. Ridiculously small bread rations were allotted to the Jews. Their purpose was not only to starve but to shame and exhaust us.*[84]

A teenager, also murdered by the Nazis, left impressions of his first day in the Vilna ghetto: *Here is the ghetto gate. I feel that I have been robbed, my freedom is being robbed from me. . . . Additional Jews keep swarming in constantly. Beside the four of us there are eleven persons in the room. The room is a dirty and stuffy one. It is crowded. The first ghetto night. We lie three together on two doors. I do not sleep. . . . I hear the restless breathing of people with whom I have been suddenly thrown together, people who just like me have suddenly been uprooted from their homes.*[85]

These observations refer only to the beginnings. A glimpse of the progressively deteriorating conditions is captured by the first diarist: *Over three hundred of us died each day in the ghetto even before they began to load us on the wagons which were to take us to the slaughter. Disease and pestilence of every kind wrought havoc among us, but hunger claimed the greatest number of dead.*[86]

With time, as the Nazis continued to deport large segments of ghetto populations, they herded the remaining Jews into fewer and fewer centers. As a result, even though the number of ghettoes shrank, the density of each existing ghetto increased. But no matter how unbearable the over-crowding, the Nazis kept moving more and more refugees into these Jewish quarters.

Coming into a strange environment, penniless and without the possibility of earning a living, many of these newcomers joined the already overflowing ranks of beggars. Beggars were the most likely candidates for death by starvation. Indeed, most of them died before they were deported to concentration camps. A disproportionate number of these beggars were children, without homes, without parents, and without hope.[87]

Years later, in 1978 in Warsaw, the rescuer Piotr Wrona could not erase the memory of these starving children. Piotr, an engineer, had worked as a locksmith in the Warsaw ghetto. Each morning as he entered the backyard of his workshop he had a similar experience: *I could not move backwards or forwards, I was all surrounded and they all begged me with cries, imploring me for food. They were not just begging, imploring . . . . The hands that were turned towards me were not hands, they were bones, simply bones covered with skin. You saw nothing, only shining eyes and faces that were already completely dull without expression. These children were kept in a church, they looked inhuman, they looked like ghosts. . . . You know what it is when a human being is starved? One is capable of doing anything! These were helpless children who were swarming me in the hope of receiving something to eat. So I would bring five rolls. I could not pass through the gate with more. The Germans would not let me.*[88]

Children, the future of European Jewry, became special targets of Nazi destruction, and Jewish leaders were acutely aware of the need to give them special protection. Extraordinary efforts to save them were set into motion. Some had to do with the collection of special funds through theatricals and other charitable events. Some of these resulted in the establishment of ghetto orphanages, others were a part of efforts to place Jewish children in Christian orphanages and convents.[89]

To be sure, attempts to stay alive within the ghettoes were not limited to children alone. Community leaders were also engaged in desperate efforts to reduce the overall death rate. They organized different kinds of fund-raising events, lectures, theatrical performances, and contests. They imposed taxes on the few who still had money. With these funds they established soup kitchens for the destitute and bought medications to combat the spread of epidemics.[90]

Some tried to overcome starvation by smuggling food from the Christian side, risking their lives in the process. Because smuggling often required movement through small spaces and because children were thought of as less conspicuous, many of the smugglers were very young.[91]

In the end, however, when faced with the determination of the overpowering enemy, all such efforts were reduced to ineffectual gestures. Of the Jews in Poland, 20 percent died in the ghettoes.[92] This death toll continued to climb uninterruptedly.

Still the Nazis continued to search for more efficient ways of annihilation.

Auf Grund des § 5 Abs. 1 des Erlasses des Führers vom 12. **Oktober 1939** (Reichsgesetzbl. I S. 2077) verordne ich:

Artikel 1.

In der Verordnung über Aufenthaltsbeschränkungen im Generalgouvernement vom 13. September 1940 (VBlGG.). I.S. 288) mit den Änderungen der zweiten Verordnung über Aufenthaltsbeschränkungen im Generalgouvernement vom 29. April 1941 (VBlGG. I.S. 274) wird nach § 4 a folgender § 4 b eingefügt:

§ 4 b.

(1) Juden, die den ihnen zugewiesenen Wohnbezirk unbefugt verlassen, werden mit dem Tode bestraft. Die gleiche Strafe trifft Personen, die solchen Juden wissentlich Unterschlupf gewähren.

(2) Anstifter und Gehilfen werden wie der Täter, die versuchte Tat wird wie die vollendete bestraft. In leichteren Fällen kann auf Zuchthaus oder Gefängnis erkannt werden.

(3) Die Aburteilung erfolgt durch die Sondergerichte.

Artikel 2.

Diese Verordnung tritt am Tage der Verkündung in Kraft.

Der Generalgouverneur
Frank

A decree issued on October 15, 1941 by Governor Frank in Poland specifying the death sentence for any Jews who moved outside of the ghetto without authorization. The same death sentence applied to Christian Poles who knowingly helped the Jews hide or move out of the ghetto.

Translation

Pursuant to § 5, Paragraph 1, of the Führer's Decree of October 12, 1939 (*Reichsgesetzblatt* I, p. 2077), I issue this ordinance:

ARTICLE 1

In the Decree on Restrictions of Residence in the Generalgouvernement September 13, 1940 (*VBlGG* [Verordnungsblatt Generalgouvernement] I, p. 288), as modified by the Second Decree on Limitations of Residence in the Generalgouvernement, of April 29, 1941 (*VBlGG*, p. 274), the following § 4b is inserted after § 4a:

§ 4b

(1) Jews who, without authorization, leave the residential district to which they have been assigned will be punished by death. The same punishment applies to persons who knowingly provide hiding places for such Jews.

(2) Abettors and accomplices will be punished in the same way as the perpetrator, and an attempted act in the same way as an accomplished one. In less serious cases the sentence may involve penal servitude, or imprisonment.

(3) Cases will be judged by the Special Courts.

ARTICLE 2

This decree takes effect on the day of promulgation.

Governor-General
Frank

In 1941, in addition to expanding and modernizing existing concentration camps, the Nazis began hastily to erect new killing centers, specifically designed for Jewish extermination.[93]

The year 1942 signaled the readiness of the Nazi death machinery, the beginning of the end for most Polish Jews. In the summer of 1942 the head of the Judenrat in Warsaw, Adam Czerniakow, was asked to deliver 10,000 Jews per day. When he realized what was happening he committed suicide. But the Nazis had not left themselves dependent on Jewish cooperation, and deportations to the killing centers proceeded on schedule.[94]

By April 1943, during the Warsaw ghetto uprising, the Jewish population had dwindled from 500,000 to 35,000. By 1944 more than 90 percent of Polish Jewry had perished.[95]

Among the many measures aimed at Jewish extermination was a 1941 decree that made any unauthorized move out of a ghetto a crime punishable by death. The same punishment applied to Poles who were helping Jews move into the forbidden Christian world, the so-called Aryan side.[96] This law was strongly enforced, and executions of Christians and Jews followed.[97] But as the anti-Jewish abuses multiplied, so did the efforts to escape from them. Grave dangers did not prevent some Jews from moving to the forbidden Christian world and some Poles from helping them to do so.

In Poland the year 1941 marks the appearance of righteous Christians. Like anywhere else these selfless rescue attempts expressed opposition to the Nazi measures of Jewish annihilation.

Jewish efforts to survive among Christians were also a reaction to the Final Solution. To succeed, the Jews' desire to save themselves by passing required aid. Both the selfless aid from Poles and the Jewish desire to pass were highly interrelated. An understanding of one inevitably adds to the understanding of the other. To gain insight into the righteous Christians and their protection of Jews I will devote some attention to the issue of passing. The next two chapters then ask: What did it mean to move to the forbidden Christian world? What obstacles did the Jews have to overcome to pass as Christians?

PART I

The World of the Rescuer and the Rescued

CHAPTER 1

Defying the Final Solution: Moving to the Christian Side

Large, wondering, dreamy yet penetrating eyes dominated her face. A well-shaped mouth with a playful smile offered a glimpse, a promise of straight white teeth. The rest too, the nose, the chin and cheeks, seemed in perfect harmony as if chiseled out of marble by skillful hands. All those well-fitting features were framed by soft, wavy, golden hair. Wondering whose photograph this was, I heard Rina Ratner's voice: *Yes, I was beautiful. In a way, my looks saved me.*

Rina Ratner came from my hometown, Lublin. During the war she passed as a Christian and then settled in Israel. When I met Rina she was in her sixties. Soft-spoken, with an air of resigned sadness, she did not resemble the spirited and confident young woman in the photograph. During the war, like most Polish Jews, Rina was forced into a ghetto along with her husband, a physician. But because she was employed in a factory located outside the ghetto, she was given a special permit to leave the ghetto at specified times. On her way to and from work she would visit a Polish working class family, the Lasows. Eventually a friendship developed. When Edna Lasow was eager to terminate an unwanted pregnancy, Dr. Ratner sneaked out of the ghetto to perform the abortion. In turn, the Lasows, although poor, shared with their Jewish friends their meager food supplies.

By 1942, it was hard to deny that the Nazis were bent on destroying the Jews. Edna and Tadek Lasow were worried about the safety of their Jewish friends and suggested that the Ratners save themselves with the Lasows' help by passing as Christians. The Lasows pointed to the advantages of their friends' Polish looks and their command of the Polish language. They even offered to arrange false papers and temporary shelter for them. But the Ratners were not yet ready to make such a move and agreed only that at the appropriate time, they would slip out of the ghetto and accept their friends' offer.

As it turned out, it was the Nazis who decided when that time would

come. In November 1942, without warning, they moved to liquidate the
Lublin ghetto, Majdan Tatarski. Many of the ghetto inmates were simply
killed; most of those that were not immediately shot were deported to a
nearby concentration camp: Majdanek. When it was all over, only a handful of Jews managed to survive, mostly by hiding in attics, closets, base-
ments, any place they could find. Among these few were the Ratners.

But, alas, the Nazis were not yet finished with Majdan Tatarski. Aware
that they had missed some victims, they tried to induce the few remaining
Jews to come forth voluntarily by announcing that nothing would happen to
those who gave themselves up within the next twenty-four hours. After
that, they said, a second search would be conducted and all Jews found
would be shot on the spot.

Although this announcement told the Ratners about the impending
danger, it also gave them a semblance of freedom. Frantically, they moved
through the ghetto trying to locate friends, choosing as their first stop the
hospital. There they found the patients lying dead in bed. All had been
shot. No one had bothered to remove them. The Ratners continued to look
for signs of life but found none. Despondent, they next tried a friend's
house, only to find another ghastly scene. Their doctor friend, his wife, and
their two children were sprawled all over the room, having escaped the
Nazis only by poisoning themselves.

Still searching, the Ratners met a few other Jews who also were wandering about, bewildered. Trapped in this graveyard, they looked to each
other for answers that none had. Rina's eyes acquired a misty distant
expression as she revisited that fateful time:

*In the past there was one particular German guard who had paid a lot
of attention to me. He was a known murderer, who killed with ease. At a
smallest provocation, or no provocation at all, he would simply shoot
whenever he felt like it. Every time he saw me come from or go to work,
he tried to talk to me. I was sure that he wanted to kill me. On that last
day as I walked through the ghetto I saw this German stagger, visibly
drunk. Suddenly when he noticed me he became furious and screamed:
"What are you doing here?" . . . Sometimes I would sleep in the factory
and not return to the ghetto. So I explained that I had come back to be
with my husband. He began to act like a maniac, repeating again and
again: "What will I do? What will I do with you?" I told him: "Do what
you want." I was already fed up. I had had enough. He said: "No, I want
you to leave the ghetto." But how could I get out? This was night, the
lights were beaming—there were guards all around, there was absolutely
no way. I shook my head. He said: "Come early to meet me near the gate.
I will take you out." I told him that I would go only with my husband. He
said "No, only you." But when I repeated that I would not leave alone he
agreed that my husband could come too. I was sure that the next morning
he would not remember any of this and that we could not trust him.
Anyway, he looked half-crazed, drunk, running around with his gun in
hand . . . people warned me that he probably intended to kill me. I felt*

that it was better to be shot right away. My husband was reluctant to come with me, but in the morning he came. The German was already waiting. He took me by the hand. Instinctively I moved away. He said: "Why do you do this?" So I asked: "After what happened you are surprised?" He told me, "I want to save you." I told him that I did not believe him. He kept saying, "I do not want to kill you, you are too beautiful to die." He then took us to the gate. The Ukrainian guards did not want to let us out. When, in anger, the German took out his gun and pointed at the Ukrainians, they opened the gate. He came with us through the gate and said: "I am leaving you here." I was afraid to turn. But I did. He stood there and said, "God be with you."

Slowly, cautiously, trying to hide from the view of others, the Ratners reached their Polish friends, the Lasows. What happened next?: *They took us right in. They were happy to see us, they worried about us. But they were also afraid. We ate and washed. In the morning Edna told us to leave and return unobserved in the evening. There were searches all around and particularly here near the ghetto. First we went to the place where I used to work; all was liquidated. The streets of Lublin were ghastly. Germans and Poles were feverishly looking for Jews. When we returned the Lasows told us that I could stay with them, but they were afraid to keep my husband. Tadek took my husband to his brother. But there too he could not remain for long, because the brother was married to a Ukrainian who could not be trusted. When he heard that a hospital for Jewish prisoners of war was still in existence, Tadek took my husband to this hospital. . . . I stayed in the Lasows' place from November till March.*

In March Rina left because she had to, and over the objections of her hosts. What made her leave? One day the Lasow's young son, who was unaware of Rina's identity, came home crying. A friend who lived next door had accused him of having a Jewish aunt. The friend, also a child, must have heard it from his parents. This meant danger. Rina explained: *I decided that I had to leave that evening; I did not want to endanger their lives. My friends did not want me to go. They were very upset and felt that I should stay. They did not believe that their neighbors would turn them in. But I insisted that I had no right to risk their lives.*

That same evening Tadek took Rina to the hospital where her husband was hiding. However, it became clear that here too her presence was endangering the others, so Rina decided to go to Warsaw. Her husband, who was less of a fighter, insisted on staying in the hospital. He lacked his wife's will to live. The Ratners would never be reunited.

Rina traveled to Warsaw with Tadek's mother, who wanted to shield her and help her find shelter. In Warsaw Mrs. Lasow took Rina to her niece, to whom she explained that, as the wife of a Polish officer, Rina had to hide from the authorities. The niece was as poor as the Lasows and as generous as they. For two months she shared with Rina her one-room apartment and her single bed. After two months Rina felt secure enough to move out and work.

And what happened to the Lasows? A few days after their Jewish guest left, the Gestapo came looking for her. Rina's quick departure averted disaster. It is hard to tell whether it was out of anger, embarrassment, or frustration, but the fact is that after the Gestapo left empty-handed, the Lasows' next-door neighbors stopped talking to them.

Rina Ratner belonged to the minority of Polish Jews who defied the Nazis by illegally moving to the Christian world, the so-called Aryan side.[1] In many ways, her last-minute decision to pass and her experiences while passing are illuminating both for what they tell us about how difficult it was to survive in the Christian world and for what they tell us about how hard it was for most Jews to make the decision to attempt to survive by passing.

One might ask why, for instance, the Ratners did not jump at the offer to leave the ghetto several months before its liquidation. Certainly few other passing Jews would be assured ahead of time of such friendly Christian aid. Why did the Ratners hesitate, even for a second?

While it is difficult for many of us today to deal with the fact that as late as mid-1942, some Jews still believed, or wanted to believe, that German aims were not all evil, such was often the case. The Germans were generally able to keep all but the most coldly analytical off-guard. Mixed in with Nazi brutality were always the subtle references to good intentions. Nazi rationalizations, for example, in late 1939, early 1940, for the forcible movement of the Jews to the ghettoes noted the extensive hostility that existed between Jews and Poles. The removal of Jews from their prior residences was justified on the grounds of protection. Indeed, once inside the ghetto, many Jews did feel safer. Still, many Jews clung to the false hope dangled before them that as bad as things were, they would not get worse. What this meant practically was that, initially, few Jews gave serious thought to survival outside the ghettoes. As long as it was possible to go on hoping, most did. Only when such hope was so obviously futile did the possibility of other options present themselves. For most, this occurred only after the start of the Russian-German War, after the summer of 1941.

At that point, two things happened. The Germans began systematically to seal off or closely guard those ghettoes that were not already physically enclosed. Simultaneously they also announced the 1941 decree demanding the death penalty for all Jews who left their residences in the ghettoes without permission.[2] This decree further specified that "the same punishment applies to persons who knowingly provide hiding places for Jews," and that "accomplices will be punished in the same way as the perpetrator . . . an attempted act in the same way as an accomplished one."[3]

For the Jews both this law and its strong enforcement came as a sudden and unexpected blow: *The execution of eight Jews, including six women, has set all Warsaw trembling. We have gone through all kinds of experiences . . . mass executions are common. But all past experience pales in the face of the fact that eight people were shot to death for crossing the threshold of the Ghetto . . . the death penalty for leaving the ghetto is completely unprecedented.*[4] But this was only the beginning. Widely publi-

---

O B W I E S Z C Z E N I E

Za nieuprawnione opuszczenie dzielnicy żydowskiej w Warszawie zostali Żydzi.

Rywka Kligerman	Motek Fiszbaum
Sala Pasztejn	Fajga Margules
Josek Pajkus	Dwojra Rozenberg
Luba Gac	Chana Zajdenwach

wyrokiem Sądu Specjalnego w Warszawie z dnia 12 listopada 1941 r. skazani na śmierć.

Wyrok został wykonany dnia 17 listopada 1941 r.

Auerswald

A poster announcing the death of eight Jews who, without permission, moved out of the ghetto. They were executed on the 17th of November, 1941.

cized, this law fast became a well-established reality and transgressions were promptly followed by executions that were also widely publicized.[5] With time the search for escaping Jews and their Polish rescuers was stepped up, and the terror became greater and greater. As the pace of Jewish extermination increased, so did the need to pass into the Christian world. Neither the presence of the law nor its strong enforcement prevented all Jews from trying to pass and some Poles from helping them.

Figures about the number of Jews who tried to survive this way are elusive. The overall estimates of Jewish survivors in Poland—ranging from 50,000 to 100,000, mentioned earlier—include those who came back from concentration camps as well as partisan fighters.[6] On the other hand, one prominent historian is of the opinion that of the 100,000 Jews who tried to live on the Polish side, 20,000 perished,[7] an estimate that would leave 80,000 who survived by passing. Most experts find this estimate too optimistic, putting the figure more realistically at 40,000 to 60,000. However, some conclude that these estimates are highly inaccurate, because after the war a large number might have continued to deny their Jewish heritage.[8]

For those who did make this move, what did the decision entail? What considerations entered into such a move?

Reaching the Christian world meant an illegal departure either from the ghetto, or a last-minute escape during a deportation from a just-liquidated ghetto, an escape from a transport headed from the ghetto to a concentration camp, or an escape from a concentration camp. Because of Nazi vigilance, runaways from concentration camps were rare. Of the 308 Jewish survivors I studied, all of whom lived illegally among Poles, only 5 percent were former concentration camp inmates.[9]

One accomplished an escape in a variety of ways. For those still in the ghetto, it involved bribing ghetto guards, with or without outside aid from

Poles. Others succeeded in smuggling themselves out on their own through secret passages and in different ingenious ways. Taking the highest risk were those who slipped away during a deportation or while en route to or from a concentration camp. Many who tried to reach the other side this way died in the process.

The mode of transition to the non-Jewish world was never certain, either for those who had definite plans for leaving or for those who were caught totally unprepared. The Nazis were careful to keep secret all dates for the implementation of their policies. Suddenly and without warning, they would add extra guards, surround a ghetto, and then proceed to round up part or all of its inhabitants. This was a frequent occurrence, and always there were some Jews who had departure plans but now had to act hastily. Many were caught even before they realized what was happening.

What factors influenced a person's decision to move and who was most likely to contemplate such a move?

The Nazis adhered to the principle of collective responsibility. Family members and Jews in general were punished for each others' transgressions. Thus if an illegal departure were discovered, others could pay with their lives. The Nazis kept exact records, so the possibility of subsequent executions was real. The danger, of course, increased when entire groups disappeared.

On a different level, escaping usually involved separation from one's family. Rarely could an entire family make such a move. Frequently one or two family members lacked the necessary attributes for staying among Poles; others had no possibility of finding refuge, or lacked the will and determination to try. Besides, it was easier and safer to slip out of the ghetto alone than in a group.

For these reasons alone, some of those who might have saved themselves did not because they could not part from their families. Ironically, strong attachments and love could and did prevent survival. At times Poles would offer to take Jewish children, particularly if they did not physically, or through their actions, betray their Jewish origin. While some parents welcomed such an offer, others simply could not face the separation. In these last instances the consequences were tragic, usually resulting in death.

Assuming, however, that a decision to pass was planned, rather than being a sudden move, and that an arrangement with prospective helpers was made, what next?

Those who planned to leave for the Christian side tried to obtain documents that would identify them as Poles. If the move involved more than one person it was safer to ask for papers with different names. In fact, Jewish families were better off pretending that they were unrelated strangers who had met only recently. If they were arrested such a story might at least spare some of them. Denying that they were members of a family diffused the risks.

Several different kinds of documents were used—some manufactured illegally and bearing fictitious names; others, duplicates of documents belonging to real people, some living and some dead.

Although it was less likely that the real documents would be detected by authorities, possessing them carried peculiar risks. It could happen that a new owner of such papers might come across the very person whose documents he or she carried, or might meet someone who knew that this individual had died. A poignant example is the true story of Ignac, a dashing, handsome young man. A newcomer to a rural village, he had no relatives and had only recently acquired friends. In no time the local people took Ignac to their hearts; the young girls in particular doted over him. Eventually one of them, the beauty of the village, fell in love with him, and the two decided to marry.

As was customary, the priest announced publicly on a few successive Sundays preceding the wedding the names of the prospective husband and wife. This was to ensure that neither was already married and that by marrying neither would be violating religious or other laws.

To a Polish woman from a nearby village the priest's announcements evoked painful memories of her dead son. Unhappy and curious, she traveled to Ignac's village to investigate. Had she approached the local priest things might have taken a different turn, but instead she went directly to the mayor, who started an official inquiry. Soon the authorities discovered the crime. The young man, a Jew, had illegally acquired the documents of the woman's dead son.

The wedding never took place. Instead, the villagers were invited to a public execution.[10]

How were such documents obtained? Some were purchased on the black market, while others were offered free of charge by different illegal organizations that smuggled them into the ghetto, when necessary. Special sections of the illegal underground manufactured, collected, and distributed false papers. Parish priests were allowed to issue duplicate birth certificates, and some of them did offer such documents to Jews. Not all Jews who lived on the Christian side obtained papers that identified them as Poles. Of the survivors I studied, fewer than half possessed such documents.[11] To be sure, some even among this group obtained their new identifications only after they had reached the forbidden side.

Most of those who didn't obtain illegal documents survived by hiding. The majority of false paper holders, however, passed as Poles without hiding. This close association between documents and hiding suggests at least two interpretations. First, possession of papers promised relative safety. Second, Jews for whom it would have been perilous to mingle with Poles made little effort to obtain such papers and instead became invisible by going into hiding.[12]

That so many Jews who reached the Christian side had no papers suggests that most of them decided at the last minute to enter the forbidden world. The idea that most passing Jews had spontaneously decided to live among Poles is also supported by the fact that 76 percent of the survivors made their move without having had Polish aid promised to them ahead of time. Moreover, even though almost by definition those who survived by passing found Poles who helped them, the majority (56 percent) were refused aid at one time or another. Finally, too, only 29 percent of survivors reached the Polish side with adequate economic resources, suggesting that their move had been sudden and had allowed little or no time for preparation. (The remaining 71 percent ranged from "moderate" to "no resources at all.")

Jews who obtained false papers had to learn many new facts to support their new identities: names, dates, places regarding not only themselves but also their fictitious relatives. Inconsistency could arouse suspicion; one slip could mean disaster.

Even perfect familiarity with supporting facts did not eliminate danger. If Jews were caught and interrogated, their captors often investigated their stories, and if the existence of "relatives" could not be proven, the captives were in trouble. On the other hand, too, a total absence of relatives would in itself lead to suspicion. It was best to have Polish friends willing to act as substitute family. In addition to offering actual or potential aid, such Poles gave an aura of legitimacy.

Becoming well acquainted with one's new identity was only a small part of what a passing Jew had to know. Familiarity with the Catholic religion was another important prerequisite to life as Poles. In Poland religious instruction extended to both home and school. Since few could escape the pressures and influences of the Church, religious ignorance was rare. Those suspected of being Jewish were subjected to rigorous cross-examinations. Failure to pass such tests often led to death. Jews who wanted to pass had to know Catholic rituals and prayers. When learning about their new religion, many benefited from the experiences of their Christian friends who directed them to the appropriate prayer books or taught them the rituals.

Some Poles were most ingenious when instructing Jews in religious customs. One of them was a woman doctor who, during the war, ran a beauty salon. A casual visitor to her place would find nothing unusual. It was filled with women wearing all kinds of creams, curlers, sitting under heating lamps and dryers. In reality, it was a school for religious instruction, operated exclusively for Jewish students.[13]

But because there was so much to know about religion, arrested Jews often had to rely on instinct to get around their limited knowledge. For example, one of the survivors I interviewed, Sara Federer, realized that the questioning was becoming too tough and that her religious knowledge was rapidly being exhausted. She began to shout that as a devout Catholic she had a right to demand a priest to whom she would answer all further questions. Taken aback, her interrogators told her that if she could bring three Polish witnesses who would swear that she was a good Catholic, she would be free to leave. Fortunately, she knew Poles who were willing to testify on her behalf. Not only was she soon released, she was also given a letter that attested to her Catholic background and asked that in the future she not be questioned about it.

Sara Federer's case, however, was rare. Most arrests involving religious tests had tragic consequences. One rescuer, Genia Parska, is still haunted by the loss of a child she had protected. During the war, as a teenager, Genia and her mother ran a boarding house in which Jews could find shelter either for a small sum or for no money at all. Mother and daughter considered it their duty to help the persecuted Jews as much as possible. Soon, however, their place became known as a haven for Jews, not only to the persecuted and needy but to the authorities as well. They were frequently visited by the Gestapo. Genia described the fate of one Jewish child and how religion was connected to it: *She was six years old. Her mother was in Auschwitz, we had her for a few months. At one of the raids by the Gestapo she was taken with us. The child was taught prayers by my mother, in case of need, in case she would be interrogated. But at the Gestapo, she refused to say the prayers. My mother begged her and begged her, yet she absolutely would not say them. Even today, I still hear my mother implore her "you know the prayer so beautifully, say it for the gentleman!" She refused and was not released. . . . I do not know how she perished. I only know that the Gestapo took her away from us.*

Who then was likely to contemplate a move to the Aryan side? Who among the Jews did in fact make such a move? Physical attributes played a part in this decision since Poles had a definite image of what a Jew looked like, and some in fact conformed to this image. Among such physically identifying features was a long nose, dark curly hair, dark eyes, and a dark complexion. In contrast, to be blue-eyed and blond, the "typically Polish look," was a definite advantage, and those with such an appearance were more receptive to the idea of changing their identity.

In reality, however, not many passing Jews seem to have conformed to this image. My evidence from two separate groups, 308 Jewish survivors and 189 righteous Poles (where such evidence is available), indicates that fewer than 20 percent of these illegal Jews had a typically Polish appearance. The majority had neutral features.

Regardless of appearance, Jewish men were in special jeopardy, since they alone in Poland were circumcised. A casual examination could reveal a suspected man's identity, and for this reason alone, passing was more dangerous for men than for women.

Was this advantageous position of women reflected in their numbers entering the Christian world? Of the survivors in my study the proportionate difference between sexes is only modest, a finding that is especially surprising when men and women are compared in terms of their Polish looks and familiarity with the language. More of the women conformed to this appearance and more of them knew the language well.[14]

To be sure, my information deals only with those who survived. Perhaps one reason a higher proportion of women did not survive by passing is that women were more likely to expose themselves to dangers, less likely to hide. Twice as many women as men worked while living among Poles. Obviously the employed were exposed to more contacts and thus to more dangers.[15] It is possible, then, that a higher proportion of men survived than would be expected because they were protected by hiding, despite the disadvantages of appearance, language problems, and circumcision.

Other ways helped too. Some Polish doctors contributed to the survival of Jews in ingenious ways, devoting their skills to changing physical attributes. A known figure among them was Dr. Felix Kanabus, who operated on noses and also devised a special operation that camouflaged circumcision. Dr. Kanabus alone was credited with seventy operations.[16] Other doctors helped by providing badly needed medical aid.

A Polish rescuer, Dr. Adam Estowski, for example, belonged to an underground organization called the "Associates of Doctors, Democrats and Socialists,"[17] which provided free medical aid to people in hiding. For security reasons these doctors operated in groups of three; each group was unaware of the other's existence. When Dr. Estowski visited Jewish patients he would also supply them with funds, allotted to them by the Council for Aid to Jews, also an underground organization. Dr. Estowski was involved in many other underground activities, not all related to Jews, but all dealing with aid to the needy. He recalled his wartime office waiting room:

I had to laugh as I opened the door. . . . Sitting there was Mrs. K. who was Jewish and passing and who came to pick up a check I had for her. Next to her was Mrs. X who also was Jewish and also had come for money. Next to her sat a Gestapo agent who had come as a patient. I knew that he was a Gestapo agent because the underground had notified me about his identity. Then there was a German who also came as a patient. Next to him was an important figure from the Polish underground who came to discuss official business with me. . . . How strange life was!

The secrecy of this medical help made it inaccessible to many, especially those who stayed in the country away from the capital.

Aside from physical attributes, the extent of assimilation also contributed to the willingness and ability to move into the Christian world. As noted earlier, in Poland Jewish assimilation was the exception rather than the rule. For centuries Poles and Jews lived apart and in different worlds. Whatever contacts there had been between them were commercial rather than social. Partly because of this, each felt like a stranger in the world of the other.

Kozłowska, a Polish intellectual who by the 1930s was already fighting the anti-Semitic excesses so common at Polish universities, continued to protect Jews throughout the Nazi occupation. She captured the essence of Polish-Jewish differences when she described a visit to her friend Sara's home in the Warsaw ghetto. Admitted to the room by Sara's father who was dressed in Chassidic clothes and who spoke almost no Polish, Kozłowska hesitated and then asked for her friend. Soon, overwhelmed and somewhat embarrassed, the two friends faced each other. *I was paralyzed*

not only by the poverty of the surroundings which made our own circum-
stances look like luxury, but more so by the unfamiliarity of the Chassidic
environment in which I was a total stranger. Suddenly I understood what a
long way Sara had to come and what terrible obstacles she had had to
overcome in order to have appeared at the Warsaw University where I first
saw her.[18]

Prior to the war, only a minority of Jews had settled in the Christian
sections of towns. These were usually the professionals, the intellectuals,
and the wealthy. Many of them aspired to assimilation, and yet of all Polish
Jews fewer than 10 percent could be described as assimilated. Actually,
neither movement out of the Jewish quarters nor a wish to assimilate
guaranteed integration into Polish society.[19] Were most Jews so different
because they refused to assimilate? Or did the Jews remain different because the non-Jews refused to accept them? Both positions have been
argued among historians. Both are extreme. The real answer may lie somewhere in between.

In wartime Poland nothing could guarantee the survival of Jews. Still,
the more assimilated were more likely to aspire to survive by living as
Poles among Poles. Of the survivors in my study about one-fifth can be
described as assimilated. But it is no surprise that Jews who survived by
passing tend to be more assimilated than the rest. Still there are degrees
and special ways in which a person can blend into his or her environment.
Assimilation of Polish Jews could be expressed by their ability to use the
Polish language. Most Polish Jews could not speak Polish well. As I noted
before, in the last census of 1931, only 12 percent of the Jews identified
Polish as their native language.[20]

But more than sheer familiarity with the Polish language was involved.
Even those who used the language in a grammatically correct way could
still be recognized by their speech. Special phrases or expressions, even if
grammatically sound, could be traced to the speaker's origin. For example,
when someone bought a new garment a Jew was likely to comment by
saying, "Tear it in good health," or "Wear it in good health." Each expression is a direct translation from the Yiddish. Also special intonations, or a
stress on certain syllables were often identifying signs. And while most of
the time Jews were unaware of their peculiar use of the language, the
listening Poles were sensitive to all such nuances.

Stefa Krakowska, one of the Christians I interviewed in Israel, told me
about her experience with language. During the war, for over two years
she hid in her house fourteen Jews, none of whom spoke Polish correctly.
Constantly exposed to a faulty way of speaking, without quite realizing
what was happening, Stefa acquired some of these habits. She worked as a
nurse in the hospital and at one point when talking to another nurse she
asked: "Nu!? Does it mean something?" Instead of an answer she received
an inquisitive look that was followed by a question: "Are you Jewish?"
Fortunately, her denial was accepted and the matter dropped. Such a small
slip, however, could have had serious repercussions.

Actually, only a minority of those who survived among Poles used out-right faulty language.[21] No doubt, too, those whose Polish was poor had to rely more heavily on Christian aid. Most of them went into hiding. On the other hand, even though many of these "pretend" Poles could be recognized by their speech, their grasp of the language was better than that of most other Jews.

But language was only one of the obvious impediments to successful passing. Overall cultural differences were among other serious drawbacks. These differences permeated all aspects of life, including such matters as eating and drinking. Thus, for example, onion and garlic were defined as Jewish foods. For anyone who wanted to pass, it was safest to profess a dislike for both. Personally, not even once did I admit to liking either: one could not be careful enough. Also, when a man refused to drink liquor he came under suspicion. Jewish men were faced with a real dilemma. If they did not drink they would be suspect. But unused to liquor, if drunk they could easily lose control and reveal some secrets. There was peril in either decision. Those who chose to drink had to make superhuman efforts to stay conscious and in control of their senses.

The survivor, Józek Braun, illustrates this problem. He was about eighteen when he moved from the Warsaw ghetto to the other side. He was protected by his good Polish papers, his flawless Polish, and his looks. He worked in a bakery where no one knew his origin. One day his co-workers, simple uneducated youths, decided to have a Christmas party. Braun who was not used to drinking hard liquor, knew that a party meant getting drunk. Each of them had to bring a bottle of vodka. Józek describes his dilemma: *I, as the others, brought a bottle of vodka. But, I never drank before. I knew I had to drink. I knew that if I did not they would begin to suspect me. At the same time I was scared of what would happen to me when drunk. Will I say something which would give me away? Still I had to drink and I got drunk. I remember sitting under the table, almost paralyzed from the effects of alcohol, unable to move at all. I also remember being aware of what I was saying. I revealed nothing, just kept agreeing with their statements. And as they were getting loud and rowdy, I imitated them, still saying nothing of importance. When the conversation turned to Jews I also agreed. This way I passed the test.*

But cultural differences extended beyond drinking. When, as it often happened, passing Jews lived among lower class Poles they had to make a conscious effort to use rough language. Those who spoke softly and politely stood out within this kind of an environment. In those days being different always suggested being Jewish—a dangerous possibility. Some succeeded remarkably well in modifying their behavior. A few of the survivors I interviewed even retained their acquired swearing and cursing patterns.

What kind of a person would be most likely to contemplate a move to the Aryan side? Once there who would have a better chance of staying alive in the forbidden Christian world? Ideally, it would be someone young, without family ties, someone who spoke Polish without a trace of Jewishness, someone assimilated who felt at ease with Poles, someone whose physical appearance conformed to the Polish look.

How does this image compare to the profile of the survivors in my study? And how do these survivors, in turn, compare to Polish Jews in general?

To recall, though far from being fluent, these make-believe Christians knew the Polish language better than the majority of Jews. Only a few of them conformed to the Semitic look. They were all relatively young at the time, most of them between the ages of twenty and forty. The majority saw themselves as agnostic and only a small minority described themselves as religious.[22] Indeed, my group of survivors included only two rabbis—orthodox rabbis, because no other form of Judaism was accepted by the Polish Jews.

Most passing Jews came from relatively high-level socioeconomic backgrounds. Well educated, most of them were professionals or well-to-do businessmen. Though financially well off before the war and perhaps at the very beginning, when they lived on the other side their economic resources were limited.

This composite profile suggests that the majority of Polish Jews—that is, the religious, the uneducated, those with limited incomes, the laborers and tradesmen—were unlikely to pass. Allowing for exceptions, the ones who ventured into the unknown Christian world were those who "fit" better into this world than those who stayed behind.

No doubt both those who did and those who did not contemplate a move to the Polish side knew that life among Poles was filled with serious obstacles and perilous risks. They also knew that ultimately the Germans were responsible for the very existence of such obstacles and perils. At the same time, they had to know that to some extent the effectiveness of these perils and obstacles depended on Poles, Poles who were eager to save and Poles who were eager to destroy.

CHAPTER 2

===

Separating Friend from Foe:
The Experience
of Passing and Hiding

One of the profound ironies of the Holocaust was that Hitler's success in stoking virulent anti-Semitic feelings in both Germany and Poland was based on the emotional argument that the Jews of Europe were not simply another ethnic minority, but rather a separate race, with separate and readily distinguishable values and, in particular, physical characteristics. Yet belying this myth was the fact that the Germans occupying Poland could not, by employing their own distinctions, separate Jew from Christian. There were some Polish Jews who looked and acted like Poles, and many more who though dark-haired and dark-eyed by no means fit the German stereotype. Unfamiliar with Polish culture, the Germans enlisted the aid of some Poles who were willing to help them ferret out passing Jews.

Who were these Poles who volunteered to find Jews for the Nazis? Polish patriotism, which was strong throughout the country, should have worked against any collaborations with the Germans. To an extent it did: there is no evidence, for example, that any Polish leaders collaborated with the Nazis, either in the rounding up of Jews or in any other way. There was not, for example, on the political level, a Polish Quisling. Similarly, I know of no Polish concentration camp guard. The situation for most Poles might well have been described as one of purposeful uninvolvement. Struggling themselves to survive, they neither aided the Germans nor hindered them. However, there were those Poles who did help the Germans, often with a perverse degree of enthusiasm, particularly in turning in Jews, and they often showed these Jews the same lack of mercy that the Germans showed. Some collaborated for the monetary rewards the Nazis made available; others, in part, because that was the law and they obeyed it. But for some the traditionally strong Polish anti-Semitism might have been the key factor. Well aware of the Polish anti-Semitism, the Germans invested much time and effort in keeping it alive.

Even as late as 1943, in a letter to a high-ranking Nazi official, Heinrich

Himmler urged the wide circulation in Poland of stories about Jewish ritual slaughter of Christians.[1] This came after the Nazis had already widely distributed posters and movies depicting the Jew as evil, ugly, dangerous, dirty, and disease-ridden. Not only had the Jews been characterized as subhuman, as vermin to be exterminated, they had also been blamed for every conceivable ill, including the war and Communism. As part of this campaign the presence of runaway Jews was constantly emphasized, and the Nazis urged Christians to deliver the fugitives. The many pressures to apprehend Jews created a virtual witch-hunt. Suspicion was rife. Anyone who lived alone, had no relatives, no friends, received no mail, was automatically suspect.

The Nazis were efficient. In addition to enlisting the necessary cooperation, they offered special rewards to anyone who would denounce a Jewish fugitive. The nature of these rewards varied depending on the locality and the demands for certain goods. These might have included rye, sugar, vodka, cigarettes, or clothing, or in some instances half the property of the apprehended Jew.[2]

Some Poles were lured by the rewards; some came under the influence of the Nazi propaganda, absorbing the definition of Jews as subhuman.[3] Ashamed of his fellow citizens' treatment of the Jews, a prominent Polish doctor observed: *In general towards the Jews there exists some wild animal-like response. A certain psychosis took hold of the Polish people, who by following the example of the Germans do not see a human being in the Jews. Instead, they perceive the Jews as dangerous and threatening animals; creatures which ought to be exterminated in any way possible just like one needs to exterminate rats with pesticide.*[4]

This observation is echoed by a Polish teacher who relates how, on stopping in a village inn, he observed that local peasants were preparing for a Jew hunt. All were in a festive mood, armed with rakes, sticks, shovels, and axes. As they drank, their jokes about Jews became more frequent and their laughter louder. No one objected to the forthcoming event; all seemed united in a common purpose. Shocked by these preparations, the teacher asked: *How much will you be paid per Jew?* An uneasy silence fell over those present. *Then I told them that for Christ they paid thirty silver pieces. You should demand the same payment.* No one answered him. Later, shots rang out and he knew a raid was in progress.[5]

Ironically, the more barriers the Jews encountered in passing, the greater was their dependence on Polish help. In a real sense, then, passing Jews were caught between two extremes: the good and the evil. To stay alive they had to learn how to keep away from prospective Polish denouncers while recognizing potential Polish rescuers. What did they have to face as they tried to differentiate between friend and foe?

Frida Nordau, whom I interviewed in New York, remembered dealing with Poles who were avid anti-Semites, determined to destroy her, and others who were exceptionally noble and ready to help. At one stage after a few futile attempts to find shelter, Frida and her family built a special bunker in the forest consisting of one hole camouflaged by branches with another hole, their actual hiding place, leading into it. The idea was that, coming across an empty hole, searchers would conclude that its occupants had fled and not look further. From their abode in this second hole, a few of them would venture under cover of darkness into the village, where local peasants gave them food. Frida remembered:

One night those who left returned with a warning that the Germans were preparing to raid the forest. Early next morning my father insisted that my sister and I leave. He wanted at least some of us to survive, since it seemed that here in the forest only death awaited us. He himself was afraid to leave; he had a beard and he knew that he could not pass for a Pole. Actually neither could we; we could be recognized through our speech and partly by our looks. . . . The moment we moved out of our hiding place Poles caught us. They were armed with axes and rakes. We begged them to let us go but they would not listen, all they said was: "You are Jewish and you have no right to live." They took us to a nearby village where they left us with the Germans. Maybe these were the Germans who were supposed to make the raid, I don't know. But after we were left there the Germans started to kid us about how pretty we were, they seemed so relaxed. I was fourteen, my sister was nine. I said to her: "Come let us walk away, let them kill us from the back. I don't want to be shot at from the front." I took my sister by the hand and we started walking away. As we did I waited for the bullets. I walked and walked and nothing happened. Then I began to think that maybe this was death. I was afraid to look back, I just walked. Later, curious, I slowly turned my head. The Germans were far away. They had not killed us. . . . I had an address of some peasants my father had given me. I decided we should go there, and maybe they would take us. Father always used to help these peasants, lending them money, doing favors for them. Now we told them our story and how we had saved ourselves. The peasant said to us: "You go into the barn and I know about nothing." It was cold, and snow was falling. We were glad; we hid in the straw. Then we heard shooting, and we saw that the forest was on fire. They were looking for Jews. I cried all day long. My mother, father, brothers were there. We had no food but we were thankful to have a roof over our head.

The Germans and Poles who conducted the raid did not discover Frida's family because seeing the first empty hole, they assumed that the Jews had left. That night the entire family was reunited. However, this was just one of many close calls. Eventually, only Frida and one of her brothers would survive.

Close calls were a common occurrence. Another survivor story, Fela Steinberg's, illustrates the need for vigilance. Fela Steinberg lived in the part of Poland close to the Russian border. Right after the Germans invaded Fela's area, they murdered a large part of the Jewish population; the rest they forced into a ghetto. Fela's father, a resourceful man, had been planning for the future. He had hired two Polish brothers to build a solid hiding place in the nearby forest. Even before it was completed they moved some of their belongings as well as food provisions into this hiding place. The Steinbergs had also built a bunker within the ghetto for use in case of a Nazi surprise raid. The ghetto hiding place was to serve as temporary shelter until they could move to the more permanent one in the forest.

In the winter of 1943, the Germans began to distribute meat, an unheard-of luxury. Fela's father, interpreting this gesture as the beginning of the end, urged his wife, two daughters, and three sons, to hide in the ghetto bunker. But Fela's sister refused to go, claiming that she did not want to be buried alive. Because none of the others were eager to hide, the father began to doubt his own decision. In a quiet, steady voice Fela pointed out how events proved her father right:

At four o'clock in the morning the shooting started. We were caught off-guard. When my father said, "Children, this is it," I panicked and ran out, I did not know where. . . . Outside I saw my mother running, and she grabbed my hand. She was going in the direction of the guards and the ghetto wire. I tried to pull her away from there, and she started to scream that she wanted to go back to father. I could see from the outside that they were already in our house. I pulled my mother with me and we followed the crowd.

Mother and daughter jumped into one of the trenches and discovered that it led into a bunker where they hid. Outside they could hear screams, feet running, and shots. After a while these noises stopped, and then they resumed. Nearby they could hear some people being dragged out forcefully, but no one discovered their bunker. When the running stopped they could hear shots close by, and then the shots came into the trenches. Mrs. Steinberg put her hand on Fela's mouth, but there was no need, for the fifteen-year-old girl was struck dumb with fear. Even after a dead silence settled in, neither of them could speak for a long time. Then the two women realized that they had shared their hiding place with a young man, Romek, whom they had known as a resourceful smuggler. The three decided to leave together in the late afternoon. At first each crawled out separately unobserved. Outside the ghetto, only at a safe distance, did they begin to search for shelter. Fela would go into a peasant's hut while the two waited outside. She would explain their predicament and ask if they could stay. They met with several refusals, some friendly and some hostile.

Finally, they came to a hut close to railroad tracks and away from the other dwellings. The Poles who lived there seemed more refined than the rest of the peasants and the three fugitives were received well. The man of the house, Stach, was a carpenter and, as it turned out, deeply involved in underground work. The two-room hut was clean and well furnished.

The two Poles, Magda and Stach, explained to the three fugitives that, since their place was exposed and the Germans stopped here often, no Jews could stay safely for more than a night. Fela's mother told them that she would turn over to them property she owned if only they would agree to shelter them. Appealing to their conscience by describing their horrible predicament, she was interrupted by Magda: *You and the boy I cannot keep, but your girl I can. She does not look Jewish.* Mrs. Steinberg began to thank her for this opportunity, but Fela angrily refused, insisting that under no condition would she be separated from her mother. Angry, Fela refused to contemplate this possible arrangement.

Soon the three went to sleep on a pile of straw spread on the floor. Fela was depressed. Lying close to her mother she did not say much. With arms wrapped tightly around her, she fell asleep. In the morning Fela woke up beside an empty space. She understood. Mrs. Steinberg had left a message with Magda that Fela should not cry and that she would try to come back. This was the last time mother and daughter met.

Compared to most other Jews who were searching for shelter, Fela was fortunate. Magda and Stach were decent people who treated their guest with consideration. Before the rest of the community Fela pretended to be an orphaned cousin who had come from the city.

Despite the relative safety, or perhaps because of it, Fela became preoccupied with the fate of her family. She missed them terribly, and the painful uncertainty developed into an obsession. Fela's hosts tried to discourage her from leaving, but she was in no mood to listen. Deciding to seek out those most likely to know something, she thought first of the two Polish brothers who had built the forest bunker. The Steinbergs had trusted them not only with the hiding place but also with some of their valuable possessions. Another person to whom Fela felt she could turn was a simple, uneducated Polish woman who had befriended her mother, supplying her with food whenever Mrs. Steinberg had ventured out of the ghetto. To this Polish woman the Steinbergs had also entrusted some of their possessions.

Fela set out at dusk on foot and stopped first to see the brothers. Their hut stood close to the forest and she reached it unobserved.

What happened next? *I knocked and when they saw me they almost died. . . . Instead of waiting to be asked, I told them where I lived. I gave them a fictitious place in a different direction from where I really was.*

Giving the wrong address was a rule all passing Jews followed scrupulously. It was a preventive measure against possible disaster. Even honorable people could be tortured into revealing the whereabouts of Jews. So engrained was this rule that most Jews preferred not even to know each others' addresses.

Fela explained that she wanted to know if any of her family had survived: *They said, of course everyone is there, father, sister, my brothers, all the others who were supposed to be there. I was so excited I wanted them to take me there. But, they told me that they would take me in the morning. In the meantime I should go to sleep. They put me in a room, closed the door, and actually locked me in and left. I fell asleep and then I dreamed about my mother. She said to me, "Run child, run!" I woke up shaking. I was locked in. I opened the window and jumped out. It was night. It was curfew, where could I go? The darkness and emptiness scared me. I de-*

cided to go to the Polish woman who was my mother's friend. She let me come in and hugged me. I told her what had happened and that these Poles had said that everyone was in the bunker. The woman became all excited: "My child, they want to kill you. No one is in the bunker. These men work with the Germans, they want to get you out of the way, so they can keep what belongs to your family. Be grateful that you were able to leave their house." I started to cry. She then let me go to sleep behind the stove with pillows and a cover.

Later on the woman consulted her husband and two teenage children about what to do next. They could not shelter Fela because, coming from these parts, she could be easily recognized. Besides, Fela was eager to get back to Magda and Stach. But travel could be dangerous, especially since Fela had run away from the two brothers. Recognizing the danger, without much prodding, the two teenagers volunteered to accompany Fela to her old place. At dawn the three drove off in a horse-drawn wagon filled with straw, with Fela lying underneath the straw. Even to these helpers Fela gave an address some distance from her real destination. Once they left her they did not even look back. They too preferred not to know. Fela remembered: *After the war I found out that no one had reached the hiding place in the forest; all were killed in the ghetto. A cousin of my father's survived and saw everything. Father ran and the Germans killed him. He did it on purpose because he did not want to be taken alive.*

Because they could be more easily identified by Poles than by Germans, passing Jews were particularly afraid of Polish denouncers. Their fears were justified, because denounced Jews were either shot on the spot or taken to the Polish police or Gestapo. It was easier to bribe a Polish policeman than a German one, but if a bribery failed, being caught by either one usually spelled disaster.

How many Jews were caught and killed because of denouncements we can never know. To be sure, the survivors in my group represent those who were successful in overcoming all dangers, yet all of them lived in the shadow of denouncements. Of the 308 survivors I studied, the overwhelming majority (88 percent) had some direct or near close calls, or at one time or another believed that their own denouncement was imminent. Often even a false alarm or an unrelated event could bring them close to death. The precariousness of the situation is illustrated by the following case.

At twenty Basia Braver lived in the Warsaw ghetto with her mother, sister, and a young niece. On the other side another sister shared an apartment with Polish friends. Pretending to be a relative of these Christians, the sister worked and moved around freely.

Basia's mother was afraid to leave the ghetto, convinced that she could only become a burden to others. Yet she urged Basia to leave, insisting that her daughter's looks and behavior promised safety. The Poles who were helping Basia's sister were willing to help her, or indeed, anyone who asked them.

After leaving the ghetto in 1941, Basia soon met and fell in love with her cousin. They then moved to their own apartment. One morning as the couple ate breakfast, strange sounds halted their conversation; any special noises could mean danger. They soon discovered that the house was sur-ιounded by Gestapo. Of the two, Basia could pass more easily for a Christian. They knew they were running out of the precious commodity, time. Basia recalled:

The Nazis were moving in . . . there was still a little time. I asked my cousin to leave, to walk by them. A Jew would not be so daring . . . I concentrated only on him. When he left I watched him from the window. They did not stop him, he passed. Then the Gestapo came to my door asking about my cousin. When I told them that he was not here they took me instead.

At Gestapo headquarters, Basia's feeling of dejection and depression was only intensified by a severe toothache. Exhausted and fed up with life, she resigned herself to whatever would happen. She had no strength to fight, nor the desire to. Drained, Basia dozed off:

I dreamed about my mother. She was standing before me . . . . The room she was in . . . was bare, airy, all white . . . not like any place I had seen before. My mother stood before me very much alive and said: "Don't be afraid. All will be well, don't give in . . . don't worry, only don't worry." "What should I do?" I screamed, but she was gone and I woke up.

Then they asked me into a special room in which there must have been fifteen Germans, conducting different business. I saw Poles come in and get money for denouncing Jews.

The Germans began to question Basia about her cousin. It was clear that someone had denounced him. They insisted that since he was Jewish she must be Jewish too. Basia denied their accusations, telling them that she had come to Warsaw from another city to escape from an unbearable family situation. To make her story more believable she gave the name and address of a Polish woman she knew in another city, claiming that this woman was her stepmother.

Though fluent in German, Basia pretended not to know any, reasoning that with an interpreter she would have more time to think about the answers. Finally, realizing that her interrogation was moving in circles, she asked to see the head of the department. *They laughed at my request. At that moment an officer came in and asked what was going on. They explained that here was this girl, probably Jewish, who had nerve to want to talk to him. He was the chief. His eyes examined me closely and he asked me to join him in his office, where the questioning started over again. I did not waver, I repeated my story; a nasty stepmother had made me leave my town, I came to Warsaw, met this man, knew nothing about him, we became lovers. . . . The German concluded his interrogation and said that I would have to go to Pawiak [Warsaw prison] and stay there until they could check out the information I gave. I told him that I did not want to go to Pawiak, that there were lice and I could get typhoid fever. So he let me go. It was a miracle. Besides, it was clear that he was attracted to me. He*

insisted that I come to his office every Tuesday until they checked out my
story. I promised to do that, but as soon as I left I ran away to the country.
They did check out my story with the woman whom I had called my
stepmother. She told them that these were lies. I found out about this from
a cousin of mine who came to Warsaw from there.

Protected by Polish peasants, Basia stayed in the country. Eventually the
mayor of the village became suspicious. Somehow Basia did not fit into this
environment, and anyone who did not fit was suspect. Basia went back to
Warsaw where she had to find another place to live. She had overcome all
the obstacles, but her cousin-lover, like so many passing Jews, had disappeared without a trace.

Blackmail was also an ever-present threat to the illegal Jew. Some Poles
practiced it as an occupation. This activity even led to a special term,
Szmalcownik (derived from the Polish *szmalec,* meaning "fat"). The Polish
underground issued a warning that all blackmailers and collaborators would
be shot, but the practice continued.[6] How many blackmailers there were
we do not know, but all passing Jews feared them and many were exploited
by them. Of the 308 Jews in my study—representing by definition success
in averting disaster—71 percent reported having experienced blackmail
(the remaining 29 percent spoke of it in relation to others).

As a rule, blackmail never ended with one encounter and it often required large payments. Frequently, too, after blackmailers had drained
their victims' resources, they would deliver them to the authorities to be
shot. One had to escape from their clutches before it was too late. A special
expression used by those who tried to pass describes an apartment visited
by blackmailers as being "burned." Burned apartments had to be avoided
at any cost. Whoever had been so victimized not only had to find new
shelter but also had to have new papers, and if employed, a new job.
Because the Nazis required every person to register with the authorities,
anyone could be easily found. Once blackmailed, a person could not simply
move to another apartment. In these cases it was best to become invisible,
at least for a while.

The constant fear of denouncement and blackmail led to tragedy. Tragedy and fear, so much a part of Jewish life, gave rise to depression, which
was often reflected in the sadness of the eyes. In the lives of passing Jews
the possibility of having sad eyes became an ever-present threat, a threat
most Jews were aware of. Jews were known for their sad eyes. They could
be recognized just by the sadness of their eyes. Many were.

Not yet ten, Pola Stein and her father were hidden by an exceptionally
kind peasant, Jan Rybak. For a short while, however, because of impending danger, she had to live with another family. In this new place, instead
of hiding, Pola moved around freely pretending to be a relative who came
for a visit. All seemed to be going well until one of the neighbors said:
Whom are you kidding, that this is your cousin? This is a Jewish child
frightened as hell, one can see it in her eyes!" Right away she had to be
sent back. It was too dangerous to stay. One never knew.

Sad eyes could become a threat for the least likely cases. Vera Ellman, an educated young woman, had what was considered to be a perfect Polish look; she was also fluent in the Polish language and there was nothing in her manner to betray her Jewish origin. Through influential friends she got an important job. An educator by profession, she became the director of a boarding house for poor women. Except for her friends who knew her as a Jew before the war, no one else knew, nor did anyone suspect. Except once. This happened in 1943 when Vera met someone from her native town who described to her in gruesome details how the Jewish remnants in her town were murdered. Many of them she knew personally; many she was close to. That day, when she returned to her place of work, her appearance must have in some way changed. One of her co-workers who had no idea about Vera's Jewishness told her that she looked sad like a Jewess. To that Vera answered: *Actually I do not know who I am. The Germans consider me as a Volksdeutch, Russians think that I am Ukrainian. You that I am Jewish. You tell me. Who am I?* She laughed. With that the incident was closed.

Sometimes, even if they possessed all the desirable qualifications, could cope with dangers, and could pass all the tests, Jews could still be recognized by Poles who knew them personally. Jews were blackmailed or denounced by acquaintances or even by so-called friends. Moses Lederman had Polish looks and could not be identified as a Jew. From his small town he moved to Warsaw where he lived with relatively few threats. Nonetheless he retains his bitterness toward the Poles, convinced that many more Jews would have survived if so many Poles had not been so eager to help the Nazis.

As Moses was boarding the train for Warsaw, he stumbled over the body of a young Jew. No one could have taken this young man for a Jew. Lederman later learned that the man had tried to leave for Warsaw but had the misfortune of being recognized by a Polish high school friend, who immediately denounced him. Without any questions a German shot him on the spot.

Similarly, David Eckstein's father was a professor of medicine at Warsaw University. He and his parents had completed final preparations for departure from the Warsaw ghetto just in time, two days before the Jewish uprising on April 17, 1943. On the other side Polish friends offered them a room and, since David and his parents neither looked nor acted Jewish, they decided it was not necessary to hide and instead lived under assumed names.

For a week, all seemed to be going well. Then one day, while they were out, the Gestapo came. With them was the Polish janitor from Warsaw University, obviously to identify the Ecksteins. Their Polish friends warned them, and they never returned. This incident was only one of many that led to threats and more moves, and the eventual split-up of the family. In the end they succeeded in overcoming all obstacles and survived.

But not all passing Jews were as fortunate, resourceful, or determined to live. As I noted earlier, Frida Nordau and her family were often forced to move from place to place. Once in the middle of winter one of their protectors, a peasant, became frightened and asked them to leave. Frida's family had to abide by his wishes. They decided that it would be safer to leave at night, and the forest seemed like the safest destination. They left on a winter night in a raging snowstorm, an unlikely night for anyone to be looking for Jewish victims. Indeed, the forest was eerie and seemed devoid of any living creatures, including men. The blizzard continued with a vengeance. The white snow contrasted sharply with the black, mysterious darkness of the night.

Into this mixture of black and white Frida's group came into a clearing where they noticed a strange object in the middle, on a tree stump. Coming closer, they realized that it was a man, half-frozen, covered with snow, motionless, and staring vacantly into space. Frida's father knew him and tried to persuade him to join them, arguing that with a group he would make it. Without changing his expression, without losing his detached bearing, the man refused: *There is no point. What will be will be.* Their pleading was ineffectual. Indifferent and resigned, he continued to sit looking into space, without seeing. They had no choice and left. As they glanced back they saw him gradually blend into the black and white surroundings. Sitting thus immobile, he took on the appearance of an inanimate object.

Another Jewish fugitive who gave up was Fela Steinberg's mother, who was guilt-ridden and depressed over having "abandoned" the rest of her family in the ghetto when she fled with Fela. She was convinced that they had all perished.

Romek, the young man who had escaped from the ghetto with Fela and her mother, became attached to the older woman during the short time they searched together for shelter. On parting he promised to look after Fela. Soon he was able to keep his promise. When he became established as a partisan, he went back for Fela.

But Magda, Fela's keeper, was fearful and cautious; when she saw Romek coming she pushed Fela into the next room, forbidding her to come out. From this room Fela overheard their conversation. Romek said he had come to take Fela. Magda replied that Fela had left long ago, and then inquired about Fela's mother. Romek explained that after he and Mrs. Steinberg parted she had found shelter with peasants, but soon had to move on to another household, and then another. Mrs. Steinberg missed her family terribly, and whenever Romek visited her she would cry. With each move she found it harder to try again. For her the whole process of finding shelter was becoming a worthless effort. Eventually Romek's pleas and encouragement lost their effectiveness. One day Mrs. Steinberg looked out a window and saw a group of Jewish men being led away by Germans. She walked out and joined them. In less than a year, for the second time, Fela mourned for her mother.

Even though the hut Fela lived in was somewhat isolated, the Poles who were protecting her could not avoid contacts with the rest of the peasants. Fela was introduced as a cousin who was visiting from town. But somehow she did not fit into this world, and people grew suspicious. After a few months rumors began to circulate that Fela might be Jewish. As the talk became more open, Magda in particular became anxious. Fela recalled:

When people around were talking and talking she really wanted me to leave. She kept saying: "I am afraid. When will it all end? You have to understand that we too want to live, you know what will happen to all of us." Stach managed to calm her down, and for a while Fela continued to stay. However, *one day three Ukrainians came on horses. When we saw them from the window Magda pushed me into the bunker, a hiding place underneath the floor. I could hear them above me. "Where is the Jewess?" . . .* Stach said, *"We have no Jewess here." The Ukrainian said they would look and if they found one the Poles knew what would happen to them. "You can search," he said. They looked around but found nothing. Before they left they said, "We will be back." After that [Magda and Stach] both told me that I had to leave. They told me to go to a Pole I knew, who they thought was keeping Jews.*

Fela left on a summer day, with little money and no documents. If questioned she was ready to say that she had lost her papers and that she had come from a village she knew had been burned down by the Ukrainians and in which practically the entire Polish population had died. In the eastern part of Poland there was great animosity between Poles and Ukrainians that often resulted in mutual murder and destruction. During the war the Nazis fueled this conflict with considerable success.

Fela described what happened: *I was feverish, I had a terrible cold. I was shot at by guards who were watching the rail tracks. The trains had been mined so many times that they watched for suspicious characters. I had to cross the rails, so they shot at me. I hid in the tall wheat and stayed there for a while. When I stood up I saw a peasant. I told him that I was going to town to see a priest because I needed a birth certificate. When he asked me where I was from I gave him the name of a village in which I knew all the Poles were killed. I told him that all my family had been killed, that I was visiting relatives, that I was looking for work. The peasant warned me that without papers I would not reach town. He took me to his hut where I met his wife and daughter. The young girl, I found out, worked for the Germans in some kind of a club. They noticed that I was sick and had fever. They must have felt sorry for me and asked me to rest. Then they remembered that their neighbor had also lost her documents and was going to town for a new copy. They suggested that after my rest I go with the neighbor.*

The sympathy shown by the peasant's family only increased Fela's sadness, making her feel sorry for herself. Exhausted, dizzy, and feverish, she lay down next to the peasant's daughter, who was resting from her night's work.

But before Fela had a chance to doze off an SS man came in. His relaxed behavior told her that he was the young girl's lover. The Nazi looked at Fela with interest and asked who she was. The Polish girl answered, "My cousin." Not fully satisfied the German turned with the same question to Fela. At that point, depleted of energy and fed up with life, Fela found this new complication to be simply more than she could bear.

With a smile I answered, "I am a Jewess." I was sure that this would be the end. I did not care. The German and the Poles thought that this was hilarious. They all burst out laughing. They thought that I was being funny. Nothing happened. The SS man left without a comment, still laughing. Later on Fela received a birth certificate and found good shelter that lasted till the end of the war. Luck, that elusive entity so much a part of Jewish survival, was on Fela's side.

The environment of the passing Jew was ever-changing and perilous. Some of the danger came from the Poles. At times even the Nazis commented on the extreme behavior of the natives. Tipped off by a Pole, an SS man remarked: *You Poles are a strange people. Nowhere in the world is there another nation which has so many heroes and so many denouncers.*[7]

CHAPTER 3

Becoming a Rescuer: Overcoming Social, Psychological, and Physical Barriers

After Germany's 1939 takeover of Poland the German administration was in full control of the country. Its subjugation of the population was by no means limited to Jews. Personal and political liberties were abolished for everyone immediately. In one of their first suppressive moves, the Germans turned to the Polish elites (i.e., intellectuals and professionals, clergy and army officers), and many of them were murdered. Others were sent to German concentration camps. Most early inmates of the just-completed Auschwitz were in fact members of the Polish elite.

However, persecution was not limited to the society's upper echelons. Following the annexation of parts of western Poland to the Reich, the Nazis began to Germanize the region. This involved removal of large segments of the native population. Such transfers were performed forcibly and without regard to human cost. As a result many Poles lost their lives.

In addition, guided by their own economic needs throughout the war the Germans continued to deport Poles for work to the Reich. Of an estimated 2.5 million who were thus used, many were worked to death, while others returned in wretched condition.[1]

Finally, too, throughout the long years of German occupation, any signs of political opposition brought a swift and brutal response. Reprisals for illegal acts led to mass killings. As Polish hostility and opposition to the occupying forces grew, so did the frequency and brutality of these reprisals. Most of those caught were executed on the spot. More often than not, these victims were uninvolved in any underground activities. The more "fortunate," those who were caught but not immediately executed, were sent to concentration camps or to forced-labor camps in Germany.[2]

Nazi abuses created an atmosphere of terror and caused many deaths. It has been claimed that the Nazis succeeded in virtually eliminating the Polish intelligentsia.[3] Other estimates put the toll in numbers of civilian victims at somewhere from 1[4] to more than 2 million[5] of a total of more than 30 million. More precise figures are elusive.

To save Jews, while their own lives were threatened, required Polish rescuers to cope with a formidable combination of physical, psychological, and social pressures and barriers. What were they?

Individual experiences of each rescuer varied in most unusual ways, but in combination they point to two broad impediments: Polish anti-Semitism and the Nazi implementation of the Final Solution.

For a concrete illustration of the rescuers' predicaments I turn first to the factory worker Stefa Dworek. At first she apologetically told me that she had not done much; she saved only one Jewish woman. Besides, it had all happened by chance.

My questions revealed that this simple but refined woman, from a poor working class background, had only an elementary school education. Mother of two sons, and practically a mother to her younger brother, Stefa Dworek had been married twice, but each husband had left her.

In the summer of 1942 Stefa's first husband, Jerzy, brought home a young Jewish woman named Irena. Ryszard Laminski, a Polish policeman who was also working for the Polish underground, had introduced Jerzy to the Jewish woman and had promised to move Irena to a more permanent shelter within the week.

Irena, who had Polish documents but could not use them, looked very Jewish, so there was simply no way she could have passed for a Christian. It was agreed that during her brief stay she would not leave the apartment. To protect her against unexpected visitors in the one-room apartment, the Dworeks pushed a free-standing wardrobe away from the wall. The space between the wardrobe and the wall became Irena's temporary hiding place. For more serious encounters they prepared a hiding place in the attic. They were able to do this because their apartment was at the very top of the stairs.

Stefa soon discovered that Laminski, a married man, was not only Irena's protector but her lover. After a week, distressed, he came to tell them that he had been unable to find a new place for his mistress. Another week passed, and then another. Laminski continued to come, each time apologizing for his failure to find Irena shelter, each time leaving food or money. But Laminski's limited funds barely covered Irena's expenses. Yet, Stefa did not mind. Although she herself was poor, she had no intention of making a business out of saving a life. They all shared what they had.

As for the awaited transfer to another haven, Laminski's efforts continued to be fruitless. After a few months, danger came from an unexpected source. Jerzy Dworek demanded that Irena leave. He insisted that his life was in danger and that he had no intention of dying for a Jewess. When Stefa pointed out that he was the one who had taken her in in the first place, Jerzy became abusive, shouting that unless Irena left he would stop coming home. In the heat of their argument, Stefa accused her husband of sexual indiscretions, while he threatened to denounce all of them. The quarrel ended with Jerzy storming out of the apartment, swearing to destroy them.

What did Stefa do? *I called Laminski . . . [and] he went to talk to my husband. He told him, "Here is my pistol; if you will denounce them you will not live more than five minutes longer. The first bullet will go into your head." After that my husband stopped coming. To Laminski he said that he could not live with a wife who has a Gestapo lover (meaning Laminski). This ended my marriage. But Ryszard Laminski continued to come, helping us, warning us about danger. He never abandoned us.*

How did Stefa feel about this development? After all, this stranger had come for one week and ended by staying for nineteen months. Was Stefa aware of the danger? *Sure I knew. Everybody knew what could happen to someone who kept Jews. I knew, but who knows, maybe I was not so fully aware. Besides, for a week I thought that it would be all right. Then it continued. . . . Sometimes when it got dangerous, Irena herself would say, "I am such a burden to you, I will leave." But I said, "Listen, until now you were here and we succeeded, so maybe now all will succeed. How can you give yourself up?" I knew that I could not let her go. The longer she was here the closer we became.*

My husband hated Jews. Maybe . . . because he did not know them. . . . Anti-Semitism was ingrained in him. Not only was he willing to burn every Jew but even the earth on which they stood. Many Poles feel the way he did. I had to be careful of the Poles.[6]

What kind of life did the two women lead? Irena never went out, and none of the neighbors knew about her. Stefa's thriteen-year-old brother was fond of Irena and helped to protect her. The fewer people knew, the safer it was. Stefa recalled: *An aunt came to me, slept here, and she did not know. Years later when she found out my aunt was angry. But I explained that she might have told someone, then that someone tell another person, and this would start a whole chain. After all, I had to be careful. It was a question of my child's and brother's life, the life of the Jewish woman and my own.*

In time the two women became good friends. Stefa recognized that conditions around them were partly responsible for their closeness. *We had to cooperate. We slept together. We ate from the same plate. I was afraid to have one extra plate on the table. If someone would come in they would see an extra plate. There was no time to hide it.*

Sometimes Stefa and her brother found themselves in strange situations. Once, for example, when Stefa was about to go to the ghetto to find out about Irena's relatives, Irena developed nervous hiccups: *A neighbor came in. My brother and I began to talk fast, sneeze, shout. We did not want her to hear the hiccups. The neighbor looked at us surprised, and left. . . . I tried to arrange it so that the neighbors would not visit me. No one suspected us. They were only suspicious when Laminski came. But they might have thought that he was my lover.*

In 1944, during the Polish uprising in Warsaw, it was too dangerous to stay in the apartment. To avoid recognition, Irena bandaged her face. In the cellar, Stefa introduced her to the neighbors as a cousin who had just arrived.

Eventually the Poles lost the battle, incurring terrible losses, and the victorious Germans began to evacuate the civilian population.[7]

Stefa and Irena heard that mothers with young children would be allowed to stay. In Irena's case evacuation would have meant exposure to many people, danger, and possible death. After all, she could not keep her face bandaged forever, and the bandage in itself could arouse suspicion. Recognizing the danger, Stefa had no intention of abandoning her friend.

When we were about to be evacuated I told Irena to take my baby. I told her: "I will try to stay with you, in case I get lost take care of him, like of your own child.". . . . When the German saw her with the child he told her to return to the apartment. Somehow I too was allowed to go with her.

Stefa cried as she spoke, but I began to doubt her story. I could not imagine giving my child away! I asked for an explanation: How could Stefa risk losing her baby? Surprised, she shrugged and then answered: *Irena would not have harmed him. She would have taken good care of him. Besides no one knew what might have become of me. I could have died too.*

Stefa continued to cry and I continued to doubt. My doubts lingered until I came across Irena's testimony before a historical commission in Warsaw. In this official document Irena praised Stefa for her many sacrifices and exceptional behavior toward her.[8] The testimony also contains Irena's version of the incident with the baby: *Before the end of the war there was a tragic moment. . . . We learned that the Germans were about to evacuate all civilians. My appearance on the streets even with my bandaged face could end tragically. Stefa decided to take a bold step which I will remember as long as I live. She gave me her baby to protect me. As she was leaving me with her child she told me that the child would save me and that after the war I would give him back to her. But in case of her death she was convinced that I would take good care of him. . . . Eventually we both stayed.*[9]

The reaction of Stefa's husband as well as the constant fear of denouncements by neighbors and relatives were related to Polish anti-Semitism—indeed, all Poles who saved Jews had to cope with obstacles that were related in some degree to the traditionally strong Polish anti-Semitism.

The cultural climate of Poland was antagonistic toward Jews. The very presence of Polish anti-Semitism implied an opposition and hostility to Jewish rescue. Those eager to save Jews were aware that by following their inclinations they would be inviting the censure of their fellow citizens.

As I have pointed out earlier, in prewar Poland in particular, anti-Jewish measures and ideologies had penetrated into religious, educational, economic, and political spheres of life. Although an integral part of the Polish culture, not all forms of anti-Semitism were explicit. The form I call diffuse cultural anti-Semitism remained vague and free-floating. In contrast to the direct and explicit anti-Jewish measures, this vague and yet all-encompassing sort of anti-Semitism has been taken for granted. It attributes to the Jew any and all negative traits, but it calls for no special action. Still, because of its pervasiveness: *Even the objectively most accomplished Jew will not be evaluated without these negative associations. One sees the Jew only through such negative glasses. One cannot free oneself of these deeply ingrained negative images. They are so common that only an exceptionally independent person can perceive a Jew as an individual to be judged dispassionately on the same basis as anyone else.*[10]

People tended to accept this form of anti-Semitism without much thought or awareness. It was expressed in such generally accepted and widely used utterances as "Be a good boy or the Jew will get you." "You are dirty like a Jew." "Don't be a calculating Jew!" and many, many others.

In the Polish language the very term Jew (Żyd) is something polite people are reluctant to use. Yet, it is the only correct term; others such as "a person of Mosaic faith" or "an Israelite" sound archaic, pompous, and downright phony. Still one can easily insult a person by simply calling him a Jew (Żyd). The term evokes strong negative images.

To this day as a Polish Jew I feel a strange sensation when I use the term Żyd. The Christians I spoke to also conveyed a certain uneasiness and embarrassment when they used the term. Thus, for example, the rescuer Eva Anielska began a story: *I had a dear, dear friend . . .* I interrupted to make sure, *Was she a Jewess?* Embarrassed, she answered: *Yes, but I resent applying this term to her.* Her friend meant much to her, and she was reluctant to refer to her as a Jew. In a similar situation when talking about his best friend, the rescuer Stach Kaminski noted: *He was my friend. I did not refer to him as a Jew. I did not see a Jew in him.*

Unobtrusive and latent though it is, this diffuse cultural form of anti-Semitism acts as an insidious foundation for all other forms. Many Poles, and particularly the rescuers, find objectionable other more explicit forms of anti-Semitism, but this almost subconscious type they tend to shrug off as insignificant. Expressions that reflect this form of anti-Semitism are dismissed as mere jokes.

Many rescuers were nevertheless conscious of their early, ever-present exposure to this diffuse cultural anti-Semitism.

Stop being a bad boy or the Jew will get you! This kind of warning was a guarantee of good behavior. *I was afraid of the Jew. I grew up with the idea that the Jew was a serious menace, a threat, to be avoided at any cost.* The speaker, Stach Kaminski, is a Pole, a highly sophisticated man, a man who has held a number of high diplomatic posts. During the war he played an important role in the Council for Aid to Jews, part of the Polish underground, devoted exclusively to helping Jews. His name appears in many historical sources of that period as one who had rescued many Jews. During the war he risked his life for the very people whom, as a child, he saw as a danger and threat. Did he come from what could commonly be described as an anti-Semitic family? Not at all. A prominent doctor, his father not only had many Jewish patients, but was proud to display his knowledge of the Yiddish language. During his childhood and adolescence Stach knew many Jews, some of whom became his friends. Yet he saw nothing wrong with ridiculing Jews and telling jokes about them, emphasizing that: *Ridicule was only an innocent pastime. It never developed into any hostile action.*

One of my closest friends was a Jew. To celebrate our high school graduation we both got drunk. It was then that my friend cried and insisted that I swear to him that I was not an anti-Semite. I remember being bewildered, not quite sure of what it was all about. . . . I knew that Jews were strange and different and that we objected to them on many grounds. But he was a friend of mine, I did not see him as a Jew. I was confused. Stach's relation to Jews, while complicated and involved, is by no means unique.

Bolesław Twardy, a known journalist-writer who spent most of his life actively fighting anti-Semitism and protecting Jews, bears a resemblance to Stach Kaminski. During the war he too was a central figure in the Council for Aid to Jews. His courageous deeds are also recorded in the history of those days. Bolesław remembers: *They tried to bring me up as an anti-Semite. As a little boy I used to run after Jews calling them names. I did this as a matter of fact, quite naturally. Once the older children made me pursue a rabbi. They taught me to call him "goy" [a term applied by Jews to Christians]. I can still see the rabbi stop, look at me with very intelligent, penetrating eyes, and then burst into free, good-natured merry laughter. His reaction made me uneasy. I felt ashamed. Later on I was less eager to run after Jews.*

Exposure to and acceptance of such general anti-Jewish views is said to have been more automatic and pervasive among the simple and uneducated. Consider the teenage daughter of a poor blacksmith who, during the war and after a major action against Jews, passed near the ghetto and noticed someone lying behind the fenced ditch. Coming closer, she realized that it was a Jewish woman, alive and in need of help. She ran for wire cutters, cut the wire, and then brought the woman to her parents, who welcomed her with open arms.

In time, a close and warm relationship developed between the family and their charge. The young girl in particular doted over the newcomer, referring to her lovingly as "my foundling." Eventually the villagers became suspicious. Danger loomed. If denounced they could all perish. But they were not about to abandon the Jewish woman. Instead, they decided that the young girl should leave with her and protect her with her typically Polish looks. As the two were forced to move from place to place, their attachment and love grew stronger. At one point the Jewish woman asked how her friend felt about Jews. The unhesitating answer was: *Oh, I hate them! The Jews are horrible. They are dirty thieves. They cheat everybody. Jews are a real menace. For Passover they catch Christian children, murder them, and use their blood for matzo.* In vain her Jewish friend tried to point out the absurdity of such accusations. The girl was only willing to concede that her friend did not commit these acts. As for the rest of the Jews, she was convinced that they did.

Frustrated and exhausted, the woman burst into tears. The young girl put her arms around her, saying: *Please don't cry, it breaks my heart to see you so unhappy. You know that you are dearer to me than a sister! But you must understand that I sucked these stories with my mother's milk. Can you expect me to give them up?* Despite her convictions about Jews, however, the blacksmith's daughter also extended help to other Jewish strangers, whom she presumably considered to be guilty of murdering Christian children![11]

The environment in which Polish rescuers lived was hostile to the Jews and unfavorable to their protection. Poles were reminded at every turn that Jews were unworthy, low creatures and that helping them was not only dangerous but also reprehensible. Not only did rescuers know that their protection of Jews would meet with Polish disapproval, but many feared that this Polish disapproval would come with actual reprisals.

Speaking to the Jews he had been hiding, a simple peasant expressed his apprehension about Poles: *I don't want anyone to know. I don't want you to leave in daylight, who knows what people might do? I am just a peasant and don't understand things, but there are bad people. You will leave at night. You will go down the trail the same way you came.*[12]

The experience of the kind peasant rescuer, Jan Rybak, also illustrates the point. Well-to-do, hard-working, and ever eager to help anyone, Jan was respected and loved by the rest of the villagers. However, when at the end of the war it became known that he had saved Jews, the other peasants underwent an abrupt change of heart and began treating their former favorite with hostility and contempt. Eventually someone denounced Jan to a Polish anti-Semitic underground group (a remnant of the wartime Polish underground that after the war refused to disband but continued to engage in terrorist activities against the Russians, the Jews, and Poles who saved Jews). Members of this group caught Jan Rybak and took him to a nearby forest to be shot. At the last moment defying his captors, Jan escaped, but after this episode death threats forced the entire Rybak family to relocate to another area where no one knew about their past.

Many rescuers, however, were not as fortunate. Wacek, for example, was a young Pole who was always cheerful and carefree, never turning away anyone who asked for help. His inexhaustible efforts on behalf of the persecuted and the needy often involved Jews. His courage and kindness in such matters earned him a reputation that eventually cost him his life. Right after the liberation Wacek was murdered by members of an illegal Polish underground group only because he had been so eager to save Jews.[13]

Polish anti-Semitism was also expressed in less severe ways. After the war when her fellow Poles became aware that Stefa Dworek had saved a Jewish woman, they considered her stupid. Similarly, Janka Polanska, young and defiant, refused to keep her help to Jews a secret. On realizing that to most Poles her protection of Jews was unacceptable, Janka began to feel like an outsider in the country she loved so well. A patriot who also participated in the Polish underground, Janka fought an inner battle. Eventually she tried to resolve the conflict by marrying a Jew and emigrating to Israel.

Janka remembered one particularly unpleasant incident that happened soon after the war. A Polish janitor who in the past had treated her with special consideration became hostile on learning that Janka had saved Jews. The furious and disappointed janitor accused her of deceiving him, because he had wrongly believed her to be a decent person. To underscore his disapproval, he returned a jar of jam that Janka had given him for his little daughter, asserting that his conscience did not allow him to accept gifts from those who saved Jews.

What about the rescuers themselves? How did they personally relate to Jews? Did they manage to escape from the pervasive anti-Semitism? Whereas most of the Polish protectors I spoke to tried to play down anti-Semitism, none fully denied its existence. Ironically, the more imbued they were with anti-Jewish images and values, the more inclined they were to say that Polish anti-Semitism was insignificant. When asked directly about their personal attitudes, all deplored anti-Semitism and condemned prewar anti-Jewish practices that dominated Poland's religious, economic, educational, and political life. In fact, some of the rescuers had defended Jews even before the war. Bolesław Twardy did it by attacking anti-Semitism in the press, a position that earned him the reputation of a "Jewish lackey." Stach Kaminski explained: *I could joke and make fun of Jews, but I opposed hurting them. When the Endeks [National Democrats] were physically attacking Jews at the university, my friends and I defended them. As a human being and as a Socialist I could not tolerate these anti-Semitic excesses.*

Eva Anielska's opposition to Jewish discrimination took another form. *I was brought up in such a way that I saw no difference between Jews and Poles. I tried to stand up for the Jews and I paid for it. Because I crossed out the word Christian on my student identification card, I was expelled from the university . . . three times. There were those who defended me and that is why I was reinstated. But later on when I completed my studies I could find no employment.*

Most rescuers consciously tried to dissociate themselves from the prevailing anti-Semitic climate and succeeded in dissociating themselves from anti-Jewish actions and ideologies. However, only a few helpers managed to steer clear of the influence of the diffuse cultural anti-Semitism. Most were caught in its clutches in different ways and in varying degrees.

Rescuers often explained that anti-Semitism existed because not only were the Jews different and strange, they were also unwilling to assimilate. In a sense, then, the Jews themselves were blamed for anti-Semitism.

The most negative view of the Jews was expressed by the Catholic writer

Marek Dunski, whose writings and political affiliation clearly identify him as an anti-Semite. At first Dunski tried to evade questions related to anti-Semitism. Only after considerable prodding did he say that he was never a "philo-Semite"—a polite way of admitting to being an anti-Semite. Eventually he said that Jewish propaganda from abroad was responsible for Polish anti-Semitism. Asked whether he thought that the Jews themselves create anti-Semitism, he answered: *There are certainly Jewish groups who do this. . . . There were tremendous differences between Jews and Poles, and these differences created resentment. It was the Jews who refused to assimilate. Maybe they thought that it was better for them to be separate from the rest of the Poles. In large measure acceptance depended on the Jews themselves. Poles accepted those who wanted to assimilate.*

Pointing to the inevitability of anti-Semitism, Roman Sadowski, a writer and rescuer, explained: *Wherever there are Jews there is anti-Semitism. People resent those who differ from them. The Jews did not blend into their environment and it was their strangeness that created anti-Semitism. In this respect, people are not different from animals. Take for example a group of ordinary mice; if you place a white one among them they will devour it. It is the same with Jews.*

While Bolesław Twardy agrees that the cultural strangeness of the Jews is responsible for anti-Semitism, he does not feel that this is inevitable. *The Jews were very different. . . . Their orthodox way of dressing in itself created a great deal of resentment. After all, their caftan was an insignificant thing, why did the Jew have to continue wearing it? It was an unnecessary source of irritation which could easily have been given up. One could dress as everyone else without giving up one's Jewishness. . . . Jews . . . are too aggressive, they do not know moderation. This has happened in Spain, in Germany, in the Middle Ages. It has always been the case if you look at their behavior historically. It is a trait which the Jews have and which has to do with their kind of upbringing . . . if aggressiveness, striving, and this pushiness to the limits would not occur, they would be looked upon positively, they would be accepted.*

Because he saw these deplorable traits as a part of the Jewish upbringing, he did not consider them inevitable.

Eva Anielska believes that the special position of the Jew in society fostered the emergence of these traits. Hers is a sympathetic view: *It is the tragedy of the Jews that they were deprived of a country. Without a homeland there is no security. The characteristics of greediness, aggressiveness, and pushiness develop in those who are deprived of a country and who because of it find themselves in an insecure and precarious position. Such traits are necessary for sheer survival. And so, I am convinced that it is the history of the Jews which is responsible for their distasteful traits. The traits for which the Jews have been hated. . . . A similar transformation took place after the 1944 Polish Warsaw uprising, when all Poles were evacuated. Without homes and without means of support, the Poles found themselves in a very insecure position. In their new surroundings, they*

*became aggressive, pushy, and greedy. They too, like the Jews, were dis-
liked by the Poles to whose places they came uninvited.*

Tomasz Jurski, a courageous protector of Jews, tried to justify and understand their alleged cowardliness: *In general, people who are constantly
discriminated against take on certain characteristics which are objectionable. Among the objectionable Jewish traits is the idea of the Jew as a
coward. But it is hard to be fearless under the conditions in which Jews
lived. The Jews were pariahs.*

As a rule the rescuers held contradictory images about Jews. For example, taking first the position that Jewish refusal to assimilate creates
anti-Semitism, the rescuer Roman Sadowski then added: *While the priest
welcomed a convert, the same was not true for the people in general. To
them the Jew was a Jew, different and therefore to be despised.* Similarly,
insisting that Jewish unwillingness to assimilate was the cause of anti-
Semitism, Stach Kaminski later said: *As far as the assimilated Jews were
concerned, we always laughed at their written and spoken Polish. Their use
of the language was faultless, too perfect, and therefore we ridiculed and
made fun of it.*

Apart from inconsistencies in the same individual, different rescuers
tended to ascribe to the Jews diverse and contradictory traits. As a group
the Jews were seen as lacking in unity, but they were also seen as too
clannish. When applied to the Jews, both of these conditions were interpreted negatively.

Stefa Krakowska, who selflessly protected Jews, felt that . . . *Jews are
too clannish. . . . A Jew would not buy from a Pole. But this was not true
for Poles, they bought from all.* Others support her position. Emil Jablonski, whose testimony I read at the Jewish Historical Institute in Warsaw,
was asked by the archivist, Jan Krupka, to meet me at the institute. I knew
beforehand that for two years he had hidden his penniless Jewish friend in
a one-room apartment and under trying circumstances. From the start
Emil impressed me as a cultured, well-informed man, and I was eager to
hear his story. When he commented about anti-Semitism he turned to
postwar Poland, asking: *When you consider 1968 and the purge against the
Jews, the Jews themselves are partly to blame. You know of course about
their meeting in Zakopane?*

No, I knew nothing about such a meeting, and urged him to tell me. *In
Poland in 1968 it was clear that the Jews had been singled out for special
treatment and that they were about to lose their power. It was then that all
the Polish Jews who held the most prominent positions called for a meeting
in Zakopane. The aim of this meeting was to devise a unified Jewish strategy, to counteract the measures the Polish government was about to take
against them. At least 2000 Jews participated in that meeting, all of whom
filled the highest posts in Poland. Naturally coming together for the purpose of opposing the existing system did not make a good impression.
Understandably such a move was interpreted as a Jewish conspiracy. Don't
you think that this show of Jewish solidarity was unwise?*

Unfamiliar with the postwar era in Poland, I asked a number of Polish historians about this meeting. They all laughed, saying that it must be a figment of someone's imagination. Not easily dissuaded, I also asked a friend of mine, a publisher of one of the most influential Polish newspapers in Warsaw. He too claimed that such a meeting took place. After a few more attempts, I had to accept their unanimous verdict.

Perhaps it is no coincidence that Emil, so convinced about a Jewish conspiracy, was also convinced that in prewar Poland all property was owned by Jews. He insisted that this was a well-established fact, reflected in a common Jewish saying: "Yours are the streets and ours are the houses."

Contrasting with the accusation of excessive Jewish solidarity was the accusation of their excessive divisiveness. The rescuer Hela Horska, for example, asserted that *the Jews are their own worst enemy. They make terrible distinctions among themselves; the Polish and Lithuanian Jews see themselves as better than the others; German Jews see themselves as superior to the rest, and so forth. There is no end to this. Every synagogue caters to different groups. It is appalling how they differentiate and argue among themselves. . . . There is no unity and no solidarity. This is a Jewish tragedy.*

Referring to his wartime experiences, Bolesław Twardy commented: *After I became a member of the Council for Aid to Jews, one of my tasks was to distribute funds and keep contact with the few remaining ghettoes. The Jews in these ghettoes were on the verge of death from starvation. Through our help we were hoping to keep them alive a little longer.*

It was then that I was again faced with this glaring divisiveness. The Jewish members of the council wanted me to report to them about the political affiliation of the Jews whom I supplied with funds. They felt that this money should be allocated only to certain political groups. To me this seemed outrageous. To ask someone who is dying of hunger what political party he belongs to was almost indecent. In good conscience, therefore, I falsified their political affiliation.

Further complicating the picture is the fact that most Christian helpers attributed to the Jews not only the negative images but also many valued and positive traits. Most admired in the Jew superior intelligence, self-discipline, hard work, and close family ties. Surprisingly, it was not unusual for the same person to see the Jew both in negative and positive terms.

Marek Dunski, who tends to blame the Jews for the existence of anti-Semitism, and who emphasizes their many negative traits, had no difficulty seeing them positively. As a writer who valued literacy, he noted that *Poles have to reach a certain level in order to read, whereas all Jews are literate. Maybe this is related to the Jewish religion. Maybe it is related to the Jewish brains. They seem more intelligent than Poles. The Jew, in difficult situations, more so than a Pole, knows how to overcome the difficulties.*

In general, a common Jew, a poor one, is more educated, more intelligent than his Polish counterpart. The Jews work hard. They like to work.

They do not mind exerting themselves. Poles, instead of working hard, systematically improvise. They lack the perseverance of the Jew. They are impatient, and want to succeed fast.

Surprisingly enough, even some Polish Jews were affected by this diffuse cultural anti-Semitism. To illustrate, Szymon Rubin, a prominent scientist and the first survivor I interviewed, was four when his parents decided to move to a small village and pass for Christians. Because of Szymon's age they felt it best not to tell him that he was a Jew. In general, he had an exceptional capacity for reconstructing the past, but flatly denied having been exposed to anti-Semitism while passing: *I heard nothing negative about Jews. After all, at that time, and in the village I lived, there were no Jews. No one talked about them. There was simply no reason to bring up the subject. . . . When the war ended two people came to visit us, two Jewish partisans. I remember wondering why my parents had dealings with Jews. They gave them food and let them stay with us. Surprised, I asked why they did it. It was then that my parents explained to me that I too was Jewish. I remember being terribly hurt. For me it was a serious blow. I was upset by this discovery.* When I pointed out that he must have been exposed to anti-Jewish views, otherwise he would not have reacted this way, he smiled in disbelief. He was totally unaware of this possibility.

Although clearly Szymon and many of the Polish rescuers were influenced by the diffuse cultural anti-Semitism, I would not label him or them as anti-Semites. Indeed, perhaps the category of anti-Semite should be reserved for a certain strength and level of adherence to anti-Semitic views rather than simple adherence per se. I do not know what that certain strength or level ought to be. What I do know, however, is that if in Poland it was so hard for some Jews to escape from these anti-Semitic influences, how can one expect non-Jews to have succeeded in doing so? Still, to the extent that Polish rescuers were at all imbued with the diffuse cultural anti-Semitism, they had to cope and overcome their own anti-Jewish images and values.

Even though Polish anti-Semitism could and did function as a collective and personal impediment to Jewish rescue, without the Nazi policies of destruction such rescue would not have been necessary in the first place. That is, while Polish anti-Semitism facilitated and contributed to Jewish annihilation, it was not responsible for it. The ultimate responsibility for the creation and implementation of the Final Solution lies with the Nazis. Moreover, their policies and actions functioned as the most powerful obstacles and barriers to Jewish protection.[14]

Foremost among these obstacles and barriers was the Nazi prohibition that carried with it the ever-present possibility of death. I have noted earlier that on October 15, 1941 the Nazis passed a law demanding the death penalty for all Jews who without permission left their residential quarters. This law also specified that "the same punishment applied to persons who knowingly provide hiding places for Jews," and that "accomplices will be punished in the same way as the perpetrator, and an attempted act in the same way as an accomplished one."[15] Any Christian who learned that a Jew was breaking this law had an obligation to report the crime or be subject to the same punishment. Determined to make both the Jews and Poles fully aware of this law, the Nazis publicized it widely; even in the most remote villages people soon knew of it. The Nazis were also determined that this law should be obeyed.[16] Transgressions were promptly followed by executions,[17] which were also widely publicized. The poster (opposite) is an example.

All Poles knew that to help Jews was to risk one's life. Frequently, however, more than their own lives were at stake. The Nazis adhered to the principle of collective responsibility. This meant that punishment applied not only to the "transgressors" but also to their families. In fact, children of all ages, including infants, were subjected to the same fate as the "guilty" adults. Nazi interpretations of collective responsibility often came to include neighborhoods, communities, and Poles in general. Equally chilling was the practice of public executions. The Germans understood that official notices of those punished were not as effective as eyewitness accounts. They were right.

One official announcement, for example, reports that on March 15, 1943, in the village of Siedlisko, not far from Cracow, the farmer Baranek and his family were executed for harboring Jews. Each name was followed by the age: forty-four, thirty-five, ten, nine, and fifty-eight. The notice ends with a statement that the four Jews sheltered by Baranek were shot as well.[18]

Years later what did the eyewitnesses remember? What did they tell about the victims and the events?

Wincenty Baranek, aged forty-four, was a prosperous and highly respected farmer. Deeply religious, he had a reputation as a generous man, eager to help others. All who knew him agreed that he was very special. Only in retrospect did some of his neighbors realize that by 1942 fear had changed him. He became quiet, less outgoing, and engrossed in his thoughts. Once, for example, this pious man went into church and stood in front of the altar without removing his hat. Only after a vigorous nudge did he remember where he was. Red-faced and confused, he apologetically reached for his hat.

His wife, Lucja, seemed to have suffered as well. She too was remembered warmly, with affection. The entire village liked and respected her. This was quite an accomplishment, almost a miracle, because Lucja was a great beauty and women ought to have been envious of her. But she never gave them any reason for gossip. Lucja was an excellent housekeeper and a wonderful mother. Her two sons, ten and nine, were model children, well mannered, and also very gifted. She was special and in a class by herself; whatever she did was exceptional. She was the only one in the entire village who borrowed books from the town's library. Every free moment she had she read.

A neighbor's glimpse of Lucja shortly before the tragedy points to the strain she must have been under. *One evening as I was passing the Baranek's house my ears caught the sound of suppressed yet desperate weeping.*

BEKANNTMACHUNG!

Wegen Verbrechens nach §§ 1 und 2 der Verordnung zur Bekämpfung von Angriffen gegen das deutsche Aufbauwerk im Generalgouvernement vom 2. 10. 1943 (Verordnungsblatt für das Generalgouvernement Nr. 82 S. 589) wurden vom Standgericht beim Kommandeur der Sicherheitspolizei und des SD für den Distrikt Galizien zum Tode verurteilt:

1. KUZYSZYS PAUL, aus Meini, wegen Teilnahme an verbotenen Organisationen,

2. KRYWEN MICHAEL, aus Meini, wegen Teilnahme an verbotenen Organisationen,

3. MROCZKOWSKI MICHAEL, aus Danylcze, wegen Teilnahme an verbotenen Organisationen,

4. BOJKO WOLODYMYR, aus Kolokolyn, wegen Teilnahme an verbotenen Organisationen,

5. KRUSZKOWSKA MARIA, geb. BOBEKOWNE, aus Lemberg, wegen Judenbeherbergung,

6. PIASTUN MICHAEL, aus Lemberg, wegen Judenbeherbergung,

7. MAZUR MARIAN, aus Lemberg, wegen Teilnahme an verbotenen Organisationen,

8. BAJEWSKI MAREK, aus Warschau, wegen Teilnahme an verbotenen Organisationen,

9. MICHALSKI WOJCIECH, aus Warschau, wegen Teilnahme an verbotenen Organisationen,

10. STRZALECKI ANDRZEJ, aus Warschau, wegen Teilnahme an verbotenen Organisationen,

11. SZUSZKIEWICZ FRANCISZEK, aus Lemberg, wegen Teilnahme an verbotenen Organisationen,

12. WOZNIAK FRANCISZEK, aus Lemberg, wegen Teilnahme an verbotenen Organisationen,

13. KANDYRAL RYCHARD, aus Lemberg-Zboiska, wegen Teilnahme an verbotenen Organisationen,

14. LEWANDOWSKI MIECZYSLAU, aus Lemberg, wegen Teilnahme an verbotenen Organisationen,

15. SAWARYN EDWARD, aus Lemberg-Zboiska, wegen Teilnahme an verbotenen Organisationen,

16. SKAPSKI KASIMIERZ, aus Lemberg, wegen Teilnahme an verbotenen Organisationen,

17. KOWALCZYK ZDZISLAW, aus Lemberg, wegen Teilnahme an verbotenen Organisationen,

18. CHROMY ZBIGNIEW, aus Lemberg, wegen Teilnahme an verbotenen Organisationen,

19. LOS BRONISLAW, aus Lemberg, wegen Teilnahme an verbotenen Organisationen,

20. BALAWENDER JOHANN, aus Lemberg, wegen Teilnahme an verbotenen Organisationen und unbefugten Waffenbesitzes,

21. KRZEMIENIECKI MARIAN, aus Lemberg, wegen Teilnahme an verbotenen Organisationen und unbefugten Waffenbesitzes,

22. NAGORNY JAN, aus Korczew, wegen unbefugten Waffenbesitzes,

23. FEDONCZYSZYN BOLESLAUS, aus Lemberg, wegen Amtsanmassung und unbefugten Waffenbesitzes,

24. PALKIEWICZ MARIAN, aus Lemberg, wegen Amtsanmassung und unbefugten Waffenbesitzes,

25. LUDALI MARIAN, aus Lemberg, wegen Amtsanmassung und unbefugten Waffenbesitzes,

26. SOLEK TADEUSZ, aus Lemberg, wegen Amtsanmassung und unbefugten Waffenbesitzes,

27. SUSCH NASTIA, geb. Diaczenko, aus Rudance, wegen Judenbeherbergung,

28. BECKER GREGOR, aus Lemberg, wegen Unterstützung von Banden u. Widerstand gegen die Staatsgewalt,

29. KONOPACKI BRONISLAW, aus Sadowa Wisznia, wegen unbefugten Waffenbesitzes,

30. HLINKIN MICHAEL, aus Lemberg, wegen unbefugten Waffenbesitzes u. Widerstand gegen die Staatsgewalt,

31. STROBNICKI STANISLAW, aus Lemberg, wegen Teilnahme an verbotenen Organisationen,

32. BERDAK KAZIMIERZ, aus Lemberg, wegen Teilnahme an verbotenen Organisationen und unbefugten Waffenbesitzes,

33. KARMAZYN STANISLAW, aus Huta Oleska, wegen unbefugten Waffenbesitzes,

34. PARSCHON PETER, aus Lany-Polskie, wegen unbefugten Waffenbesitzes,

35. LESZCZYNSKI MICHAEL, aus Ruda Sielecka, wegen unbefugten Waffenbesitzes,

36. KALINIEWICZ WASYL, aus Lany-Polskie, wegen unbefugten Waffenbesitzes,

37. IRZEK JÚLIA, aus Lemberg, wegen Judenbeherbergung,

38. MALAWSKA geb. WILCZYNSKA Wiktoria, aus Lemberg, wegen Judenbeherbergung,

39. SLADOWSKA HALINA, geb. KRZYMIENIEWSKA, aus Lemberg, wegen Judenbeherbergung,

40. JOSEFEK MARIA, geb. SLOWICZ, aus Lemberg, wegen Judenbeherbergung,

41. JOSEFEK BRONISLAW, aus Lemberg, wegen Judenbeherbergung,

42. STELMASZCZUK GEORG, aus Hohołow, wegen unbefugten Waffenbesitzes,

43. HALICKI ROMAN, aus Warschau, wegen Teilnahme an verbotenen Organisationen und unbefugten Waffenbesitzes,

44. PUKAS EDUARD, aus Lemberg, wegen Teilnahme an verbotenen Organisationen,

45. WNUK EDUARD, aus Lemberg, wegen Teilnahme an verbotenen Organisationen,

46. DZIERZEK CZESLAW, aus Lemberg, wegen Teilnahme an verbotenen Organisationen,

47. ZIEBA MIECZYSLAW, aus Lemberg, wegen Teilnahme an verbotenen Organisationen,

48. PNIEWSKI HENRYK, aus Otwock, wegen Teilnahme an verbotenen Organisationen,

49. PURSKI AUGUST, aus Lemberg, wegen Teilnahme an verbotenen Organisationen,

50. WOJCIECHOWSKI TADEUSZ, aus Lemberg, wegen Teilnahme an verbotenen Organisationen,

51. WOJCIECHOWSKA geb. DOWMOND DANUTA, aus Lemberg wegen Teilnahme an verbotenen Organisationen,

52. NUSZKOWSKI BERNHARD aus Lemberg, wegen Teilnahme an verbotenen Organisationen,

53. WOJCICKI LUDWIG, aus Lemberg, wegen Teilnahme an verbotenen Organisationen,

54. GIBALSKA ALEXANDRA EDWARDA, FRANCISZKA, aus Lemberg, wegen Teilnahme an verbotenen Organisationen,

55. SZAWEL STEFAN, aus Lemberg, wegen Teilnahme an verbotenen Organisationen,

Public Announcement!
The head of the SS and police in the district of Galicia announces that on December 14, 1943, among the 55 people who were sentenced to death for a variety of crimes, there were eight Christians who were sentenced to death for hiding Jews.

When I came closer I saw Lucja framed in an unlit window holding a handkerchief over her mouth. Her sobs were filled with pain as if at any moment they would break her heart. Amazed, I asked: "Lusia, what in the name of God happened to you my dear?" Immediately she stopped. In a casual but strange voice she said: "Oh nothing! Nothing at all. It just came over me, for no reason at all." She vanished into the darkness of the room.

When the Germans drove up to the farm at dawn, Baranek had time to ask a neighbor if they had visited others as well. Upon hearing that they came only to his place, he said: *I am lost, all is lost, the children!* In the initial confusion, caused by the arrival of the Nazis and neighbors, Baranek succeeded in telling the children to hide. They did. But as soon as the Germans realized that the boys were missing, they stopped searching for the Jews and began to look for them. After they found the boys, they shut them up in a room and ordered one of the neighbors to watch them.

Later on, people said that the children were about to jump out of the window. They might have succeeded in saving themselves, because the Germans were busy in another part of the house. But the neighbor who was standing guard forced them back into the room: *How his conscience allowed him to do this I don't know. Still, now when we think about this terrible past, we feel that we ought to have behaved differently. But then it was something else. In those days, at the Baranek's farm, we were terrorized and paralyzed into inaction. One wrong move could have meant death. Passively we watched as the Germans led out the husband and wife.*

The couple moved toward the barn, erect and stiff, as if they were to be wed. Only their faces were already dead. The atmosphere was churchlike, eerie. The silence was total, oppressive. When the two reached the barn a German soldier gave a key to the man, directing him to unlock the door. This done, the two were led inside. Then husband and wife knelt.

Still in silence, the Germans brought out the boys. The children held hands. They, too, were made to walk toward the barn. Their bodies refused to remain calm and shook vigorously. At the sight of the children the crowd burst into loud sobs and groans. But the guards reacted swiftly. Shouting *"Ruhe, Ruhe"* (quiet, quiet), they followed up by indiscriminately hitting whomever was close to them with the butts of their guns. The crowd obeyed; the noise stopped. In silence they watched as the children were made to kneel in front of their parents. Shots and more shots exploded into the heavy air. The kneeling figures tumbled in quick succession, then scattered in different directions as if taking leave. Stunned, the crowd looked on. But the executioners had not finished. They brought out four men, the Jews who had been sheltered by the dead farmer. In full view of those present, the soldiers shot each of them several times. They continued to shoot into their dead bodies, as if trying to kill them over and over again.[19]

Instead of bullets, the Germans used fire in the village of Stary Cieplow. Here, too, they came at dawn, surrounding four farms at once. Because two of the farms were next to each other, the Germans moved one of the families—husband, wife, and five children—out of their own and into the neighbor's hut, a fragile old structure in need of repairs. In each place the victims were closely watched as their most valuable possessions were loaded onto special wagons. This done, they set fire to the three houses. That day thirty-three Poles and an unknown number of Jews were burned alive. In a nearby house a niece of one of the families stood at her window. Let me reconstruct her story:

The guards led them in the direction of the neighbor's hut. The man, Adam Kowalski, at the head of the group, held the hand of each son: Henryk, six, and Stefan, five. It was like when he would go with the children for a walk, except now his head was bent down low, his legs were dragging as if refusing to move. The man was followed by his wife. In her arms she held a baby not yet seven months old. Two girls (Janka, sixteen and Zosia, twelve) walked behind their mother. Janka had her new boots on, a cherished gift from her father. The seven, surrounded by the German soldiers who kept prodding them on with the butts of their guns, moved on. When the mother reached the open door she took the baby from her chest, turned toward the soldier, and made a gesture, as if offering him the child. Perhaps she was begging for mercy? Perhaps she was trying to save the little one? The German pushed her violently. Still clinging to the child, she fell somewhere beyond the threshold. Mother and baby disappeared from view. . . . The guards shut the door securely and then encircled the hut. Against the white snow they looked like black columns. Immobile, they waited.

Soon, above the gray sky, from two different directions, came a pale pink patch. Fire! This must have been the signal the Germans were waiting for. Two of them moved closer to the hut. An explosion, a broken window were followed by desperate cries. Were these voices real, or was it an illusion? Eagerly the fire spread into all directions, up and around, and around again. With it came black thick clouds.

Suddenly the door fell. Someone ran through it. It was a girl. She ran ahead with her arms moving widely trying to push away the fumes. Her new shiny black boots contrasted sharply with the white snow. A long, thick golden pigtail, half undone, bounced against her back. Janka had escaped from the burning grave! The Germans lifted their machine guns but did not aim. Was it a miracle? Would they let fate decide and spare the girl? No; they were teasing! Only for a moment did they give hope. Hope that made what followed even more horrible. They fired several shots. The girl staggered, lost her balance. With out-stretched arms, like a wounded bird, she fell facing toward heaven. In no time two Germans were at the victim's side. Like hyenas, they came to inspect their prey. Were they still searching for signs of life? No. They were tempted by something else. Roughly, one of them pulled the boots off the dead body. Another grabbed the girl's golden pigtail and dragged her over the snowy ground. When he came close to the burning hut he lifted her up like a useless discarded sack and tossed her into the flames.

Making sure that all went according to plan, the Germans continued to stand guard. They left only when faced with a heap of half-burned ashes.[20]

Occasionally, the authorities would spare the very young. This, for example, happened to Henryk Kryszewicz Wołosymowicz. Too young to remember, he knows the family's history. During the war Henryk's parents had sheltered Jews. One night the Germans came to their farm, shot his father and the Jews. They took the mother to prison, but she never came back. Two years old, Henryk and his four young siblings were spared.[21]

At times the Nazis would show leniency even toward the guilty. The Polish socialist Płuskowski was among these few fortunate cases. A clerk in the city administration of Warsaw, he was assigned during the war to the Warsaw ghetto. There Płuskowski established contact with the Jewish underground and provided them with forged documents and ammunition. After the liquidation of the Warsaw ghetto he continued to shelter Jews. Eventually, the Nazis caught up with him and sent him to a concentration camp in Germany. He was liberated by the Americans in 1945.[22]

In sharp contrast was the fate of the young Pole, Nowak. In the spring of 1944 in Skarzysko at the age of twenty-five, he was publicly hanged in the town's square. His crime was the smuggling of food to starving Jews who lived and worked in a local ammunition factory.[23]

The wide range of punishments is further illustrated by the burning of entire villages. In the winter of 1944 the Germans learned that the village of Huta Pienacka offered food and occasional shelter to about 100 Jews from the surrounding forests. As a reprisal, the Nazis, together with the Ukrainian police, surrounded the village and set fire to it allowing no one to leave. All day long the assailants kept watch, making sure that no one left. In this case even the animals were made to share the fate of the villagers.[24]

Despite occasional and partial lapses, the principle of collective responsibility remained an awesome reality, as a rule including the rescuer's entire family. Did this principle prevent Poles with families from participating in Jewish rescue? Not likely. Most Polish helpers were married, and many had children.[25] Moreover, a majority said that in their efforts to save Jews they had the support of their families.[26]

Yet some Poles preferred to conceal their help to Jews even from those with whom they shared a home. Some rescuers were not sure they could trust their relatives; others wanted to shield their families from anxiety and possibly from death.

Staszek, a carriage maker, was among the rescuers who kept his aid activities secret from the relatives in his household. A young unmarried man, he lived with his mother and sister, who he felt would have objected to his protection of Jews. Initially, Staszek offered shelter to one close friend, who then asked that this privilege be extended to his relatives and friends. Requests for shelter from new people continued to grow but Staszek was unable to refuse, rationalizing that saving one Jew could lead to the same death penalty as saving many. Eventually, he was keeping thirty-two fugitives in a bunker he had built underneath his house.

At one point the Nazis caught him in a raid and were about to send him to Germany for labor. Knowing that the lives of thirty-two people depended on his freedom, disregarding caution, he escaped. This was only one of many close calls, but no matter what happened he was determined, and he continued to take care of his group for almost two years. Until the very end neither his mother nor his sister knew that anyone else besides Staszek shared their home. All survived. For his valor Staszek was awarded a gold medal by the Israeli government.[27]

When helping Jews, then, Poles had to overcome several layers of obstacles. The outer and strongest layer was the Nazi prohibition that made helping Jews a crime punishable by death. Next came the explicit anti-Jewish ideologies and the pervasive anti-Semitism that made help to Jews both a highly dangerous and disapproved of activity. Last, these Poles had to overcome their own diffuse cultural anti-Semitism. While struggling with these layers of impediments, what kind of aid did these Poles offer? What was involved in offering this aid?

CHAPTER 4

From Warning to Long-Term Shelter: Forms of Christian Rescue

How did the rescuer-rescued relationship begin? Did this life-threatening association involve careful planning and preparation? My evidence gives a negative answer. In most cases those in need of protection initiated the relationship by asking the Poles for help.[1] Similarly, most survivors say that help was not promised to them ahead of time.[2] Of the Polish rescuers themselves, less than one-third initiated their aid to Jews.[3] Even those accounts that deal with prolonged sheltering of Jews show that such unplanned beginnings only gradually developed into more extensive aid.

When I met Hela Horska in 1978 in Warsaw she was widowed and lived alone under modest circumstances. This was in sharp contrast to her prewar and wartime position as a nurse and wife of a prominent Polish doctor. During the war she and her husband protected fourteen Jews for over two years. How did this aid come about?

The Horskis lived close to a Jewish section of a small town. Dr. Horski was a busy physician. During the war, he was assisted by his wife. After the establishment of the ghetto, the Horskis were approached by a Jewish woman patient who begged them to employ her thirteen-year-old son David Rodman. The mother feared that the boy's delicate health would not withstand the strenuous work the Nazis demanded of him. Feeling sorry for the mother, Hela secured permission from the authorities to employ young David, arguing that her work with her husband did not leave enough time for her children and house chores. Soon the boy became an asset because of his winning personality and his hard work. Eventually he won the hearts of the entire family, including the children. Once David had become a valued member of their household, it seemed natural for his employers to want to shield him from danger. Whenever they heard about moves against Jews they warned him and hid him in their house. Eventually David felt secure enough to ask that this privilege be extended to other members of his family. Each time there was to be a deportation in the ghetto a few of his relatives would come and hide, until their number grew to fourteen. During the final liquidation of the ghetto, David's entire family remained in the Horski household. The Rodmans asked for a week's stay, and the Horskis agreed to keep them. David and his relatives expected to be smuggled into Hungary, where at that time the Jews still lived in relative safety. A week passed, but the Hungarian trip did not materialize. The Rodmans had no place to go, and the Horskis did not have the heart to send them away. Weeks turned into months, and eventually into two years, and then the end of the war.

Different, and yet similar, is Emil Jablonski's story. During the war Adam shared a one-room apartment with his wife and mother-in-law. He made a modest living as an administrator of a building. His protection of a Jewish friend, a lawyer, started as follows: *It was 1942, the gates of the building were locked, the curfew was on. I heard a hesitant knock at the door. When I looked, there was Kazik, dirty, unshaven, and sad. I rushed to cover the window so that no one would see him from the outside. Then I asked him in. He told me that the Polish woman with whom he had been staying had sent him away. He had gone to a few friends, but they had all refused to keep him. He assured me that he did not want to endanger me, he had an address to which he would go the next day. He hoped to find permanent shelter there.*

We let him take a bath, fed him, and made him a bed. Because he looked Jewish, he could go out only in the dark when people could not see him well. For his safety I decided to accompany him to the new address. When we reached the place the owner of the apartment, too scared to even talk to us, emphatically refused to shelter him. He had another lead that he followed up and that also resulted in a definite refusal. He had no more addresses. He had nowhere to go.

What could I do? I took him back to my house. I discussed it with my wife. She agreed with me that we could not let him go and die. We decided to keep him until we could find someone with a less exposed place than ours. In the meantime we moved a closet away from the wall, and placed his bed there. And this was where he stayed for over two years.

As a rule unplanned offers of help involved more than one fugitive and more than one kind of aid. A very general distinction has been made between organizational and individual kinds of rescue. Individual aid is based on the personal decision of rescuers. Organizational aid can be traced to the demands or requirements of an organization. For example, if members of an underground organization separately and on their own decided to save Jews, their aid would be classified as individual. On the other hand, if a few people in a small unit set out to protect Jews, without the knowledge and approval of the larger unit of which it was a part, this kind of help is hard to classify. Some have claimed that in Poland members of the underground were also extending protection as individuals.[4] This, indeed, is reconfirmed by my results. Of the rescuers, only a small minority limited their aid to organizational forms. More than half of them offered individualized help only, while more than one-third participated in both individualized and organizational rescue.[5] Often appearing together, individualized aid seems more common than the organizational kind.

Regardless of these and any other distinctions, receiving aid, all kinds of aid, was a virtual precondition for Jews who wanted to survive by passing or hiding among Poles. In fact, of those who did survive this way, 95 percent reported having benefited from Christian help; moreover, closer inspection of the 5 percent who claimed that they had survived on their own, revealed that they also benefited from some assistance. In their cases this help might have been limited to single encounters: the giving of false documents, being escorted out of the ghetto, or simply being offered food or lodgings for a night. Insignificant as these acts may seem, their absence more often than not could be fatal.

However, meaningful aid at times came in less tangible form. Consider the following experience.

At thirteen David Rodman's tall lean figure, stooped narrow shoulders, and pale transparent complexion revealed his physical fragility. Despite David's appearance, or perhaps because of it, he was caught with a group of Jewish men and forced into heavy labor. When David and his companions were marched toward their destination, the German soldiers impatiently urged their charges to move faster and faster, with rough voices accompanied by gun butt blows. Under the barrage of these indiscriminate beatings the men began to bleed and stagger. Since stopping could mean death, they continued to make a superhuman effort and marched on.

On their way they passed a small house. Near it stood a young Pole not more than twenty. Although bruised and hurting, David noticed him. *He had intelligent eyes and a noble face, which expressed deep sorrow and compassion. Then I heard him say: "What kind of unfortunate people are these Jews? What do they want from them?" I know exactly the house next to which he stood. I still can see the look of suffering on the young man's face, the exact color of his shirt. It did me so much good to hear him say these few words. It impressed me that someone felt for me and cared because I suffered. For a long, long time I felt less deserted.*

Remembered for forty years, the incident obviously made a deep impression. Yet, the young Pole's expression of sympathy does not fit the actual definition of help, nor does it identify him as a rescuer. Still, this Pole's reaction was effective in keeping up David's spirits. It gave him courage and hope.

All aid contained this elusive ingredient that may be identified as hope. On the more concrete level, the act of protecting a Jew took on an almost infinite variety of forms. What categories of rescue emerged from this great diversity? How pervasive were they? What do these different forms and aspects of aid mean?

Hope, this intangible ingredient of help, could and did appear at all times and places, while other more concrete forms of aid were peculiar only to some situations.

However, to become effective, help had not just to be available. Jews

who could benefit from such help had to be ready for it. For many complicated and realistic reasons many were not. This was particularly true for
ghetto and labor camp inmates, who did not contemplate a move to the
Christian side. Lack of planning in turn seemed like a natural partner for
apathy and resignation.

Some rescuers devoted their energies to helping ghetto and labor camp
inmates. (It was next to impossible to protect concentration camp inmates.)
They, more so than others, were faced with Jewish resignation. Zygmunt
Rostal was one of these rescuers.

Zygmunt's childhood before the war was one of utter poverty. His father
was a chronically unemployed laborer. The economic burdens of his family
of six children were carried by his mother, who worked as a chorewoman.
Early in life he knew real hunger, which in his estimation made him
especially sensitive to the sufferings of others. Despite the hardships, Zygmunt remembers his early years as warm and full of love. His closely knit
family was deeply involved in leftist politics. In fact, his father's inability to
find work might have been related to his political participation. As a teenager Zygmunt joined the Red Scouts, one of the few organizations that
welcomed Jews. Partly because of this, his peers contemptuously referred
to him as "Jew lover."

During the war, because of his Red Scout affiliation, he was approached
by a Socialist organizer who asked him to take a job in an ammunition
factory for the purpose of helping Jews who lived in the factory and were
treated as concentration camp inmates. As a Christian employee Zygmunt
had freedom of movement. His privileged position allowed him to supply
the Jews with food and medication and gave him a chance to try to raise
their low spirits. Zygmunt also gathered information about the Nazi moves
and intentions that affected the Jewish inmates, and supplied this information to his underground contacts.

Looking back and comparing the different kinds of aid he offered to the
inmates, Zygmunt thinks that keeping up their spirits was more important
than providing food and medication.

Zygmunt told me about a sophisticated Jewish prisoner who became his
close and special friend. The two had long philosophical discussions about
the meaning of life, God, and humanity. In addition to sharing food with
this undernourished man, Zygmunt tried to keep up his faltering spirits
with talk of a better future. Soon, however, this eager helper noticed a
marked change in the prisoner. Zygmunt's friend became progressively
more depressed and less attentive; he finally became convinced that life
was worthless. Zygmunt remembered:

*He looked very neglected, unwashed. I said to him, instead of talking let
me shave you. I brought a razor and a dish of water, but he refused and
would not let me do anything for him. He just did not care. He stopped
reacting. I warned him that one day they would finish him off. But he was
indifferent. I was helpless. And so it happened. One day he did not react to
a German's order to get out of the way . . . he was murdered while I*

looked on . . . it was horrible. I still cannot talk about it quietly. At the end of the story Zygmunt's choked voice made me pause.

Unlike his friend, Zygmunt refused to give up. Three times he was caught, three times he was beaten. Even now he suffers from a kidney condition that developed right after the third beating. But this did not stop him: *The next day I was already thinking how to go about my aid so they would not catch me.*

He continued to help until 1944 when the Nazis liquidated the place. Some of the Jews he protected survived. After the war they looked for Zygmunt, remembering his devotion. Eventually he was awarded a Yad Vashem medal.

Other Poles saved Jewish lives by warning them about forthcoming disasters. Because the Germans tried to take their victims by surprise, to warn Jews about impending dangers could be lifesaving. In fact, 61 percent of the rescuers at one time or another alerted Jews to future dangers and thereby shielded them from greater perils. Many survivors (43 percent) concur that they benefited from Polish warnings.

How a seemingly simple act of warning could have far-reaching consequences is illustrated by Bolesław Twardy's experience. Already in 1940, Bolesław had illegally transported Jewish friends from Łódź to Warsaw. This act was soon followed by Bolesław's warning to an entire Jewish community about an impending disaster. How did this happen?

One day a school friend of mine (a Pole) told me that the ghetto Brzeziny was going to be liquidated. This was a small town that made a living only from tailoring: they made suits for sale in Africa. . . . I wanted to warn them. I went in a carriage pretending at the gate that I was carrying food. I knew, because I worked for the secret service in the underground, that at a certain hour a carriage moved into the ghetto with food. . . . I had friends there. I warned them about the impending liquidation and they simply ran away. Only those that did not want to fight for their life remained, old people, sick, those who were afraid.

Warning, however, did not necessarily lead to actual escape because some of those warned would not act on the information. A move to the forbidden Christian world required extensive arrangements; bribing guards, familiarity with secret passages, and most important it meant a threatening and perilous future filled with the constant possibility of death. The presence of an experienced guide, a Pole who would take a fugitive to safety, offered a measure of security. Without guaranteeing safety, Polish escorts offered a necessary step towards freedom.

For some Poles escorting Jews from the ghetto, or from one home to another on the Christian side, was a common way of helping. In fact, an overwhelming majority of rescuers (75 percent) did at one time or another escort Jews.[6] Some even made a regular practice of it.

A single act of escorting someone to a new place was relatively brief; yet such acts often led to dangerous, unexpected complications. This is illustrated by Tomasz Jursky's experience, who as an underground worker had frequently smuggled Jews out of the Warsaw ghetto. When, on April 18, 1943, Tomasz approached the ghetto to bring two Jews to his father's house, he found the entire area surrounded by guards and the opening he usually used for slipping in and out blocked. Young and adventurous, Tomasz was not about to give up. Instead, he sneaked into an adjacent house and waited for a favorable moment. Just before the curfew he succeeded in entering the ghetto. Inside a creeping tension seemed to envelop him. Something was going to happen, it was in the air. Whatever it was, it created an overall fear and anxiety.

Many ghetto inmates knew Tomasz. Some were convinced that he was a Jew who lived on the Christian side, pretending to be a Pole. On that night his friends urged him to move out. Those he came for had already left.

But the late hour prevented Tomasz from leaving, and he decided to stay overnight. The next day was April 19, 1943, the start of the Warsaw ghetto uprising. As yet, the fighting had not spilled into Tomasz's part of the ghetto. All routes of escape, however, seemed sealed. As he searched for a way out, Tomasz met two young Jews who, like he, were eager to leave. The three promised to cooperate. Eventually the two young men managed to bribe a policeman who stood guard while Tomasz and his two companions climbed over the ghetto stone wall and into the Christian world. The fate of those left behind belongs to history; only a handful survived.

Christian aid did not end with the fugitives' arrival on the forbidden side. On the contrary, life among Poles demanded continuous aid. What kind of help did these Poles offer? What did this help entail?

For those who entered into the illegal world, shelter was of utmost importance. While indispensable from the Jewish perspective, for the Polish helpers it was dangerous and taxing. Shelter, however, was the most commonly offered kind of help. Of all the 189 rescuers, the overwhelming majority (84 percent) shared their homes with their charges. Similarly, most of the 308 survivors reported that their protectors (65 percent of 565) offered them shelter.

Not all sheltering, however, involved hiding. Of the survivors, 46 percent were hidden and another 10 percent had mixed experiences including hiding. The rest were passing as Poles. That is, they lived among Christians pretending to be like them.

Poles who hid Jews were faced with more burdens than those who simply had passing Jews share their homes. Building an appropriate hiding place that would withstand the scrutiny of unexpected Nazi raids was a major concern. Feeding the fugitives was another important requirement. Throughout the German occupation, food was scarce. Only those who were officially registered were entitled to special rationing cards. Food on the black market was expensive, but available. The majority of both Polish rescuers and survivors report giving and receiving food.[7]

The cost, however, was not the only burden involved in feeding illegal Jews. Buying and carrying large quantities of food could be dangerous if noticed by the wrong people.

Stefa Krakowska, who hid fourteen Jews, shared the food shopping with her father. They each bought in faraway places where they were unknown and also tried to shop at different times of the day. Varying the hours reduced the chance of neighbors seeing them carrying large quantities of provisions.

Those who hid large numbers of Jews frequently devised ingenious ways of purchasing and getting food into their homes, the most common of which involved rotating the shopping places, hours, and people who did the buying.

Another ever-present potential hazard related to the fugitives' health.

How does one call a doctor for someone who does not exist? Worse still, how does one bury a body that isn't there? More often than not, those in need of medical care did not know about the existence of the special underground units that offered medical aid to fugitives. The presence of such units remained a well-guarded secret both from the authorities and from most of those dependent on such aid.

The rescuers and the rescued were each aware about the disastrous turn that a fugitive's serious illness could take. And when, as inevitably happened, people became sick, neither the rescuers nor the other fugitives could avoid sharing the suffering.

David Rodman is still haunted by such memories.

Moving to a different time and place, imagine a barn in a Polish village with a small attic with a low ceiling. The attic has no toilet facilities, no water, no light. A square opening with a steep ladder leaning against it serves as the entrance and as the main pathway for air. The only other way through which air reaches the area is a small window on the side of the barn, close to the roof. Both of these openings make for poor ventilation. The attic was built as an extra storage room. Most of the time it is empty. In the summer the heat beats mercilessly upon its thin roof making the inside unbearably hot.

In the past no one was affected or worried about the poor ventilation or heat. The war changed all this. Fourteen Jews, ranging in age from three-and-a-half to sixty are brought into these cramped quarters by an old Pole who wants to save them. The straw scattered on the ground serves as their beds. A single pail becomes their toilet. Most nights their protector comes to take care of them. He brings them water, bread, and potatoes. It is not easy to feed fourteen people. It is dangerous. Here in the village people are suspicious of one another, the neighbors are inquisitive. Some nights seem especially threatening. The old man is afraid. At such times the Jews have to do without food and water. They wait.

Washing is out of the question, as is the changing of clothes. Dirt, lice, and different kinds of vermin, no one can even identify, become their constant companions. Names of the vermin don't matter. What matters is that fourteen human beings are defenseless, and at the mercy of these small and yet terrifying intruders that crawl all over them and bite. Their skin becomes infected and full of sores. Reduced to this pitiful condition some of the fugitives seriously consider giving up. Instead, they make a superhuman effort not to complain, especially not in the presence of the Pole, their savior. After all, they are among the lucky ones. They ought to be grateful. The alternative is death. As they wait and wait they cannot even decide which season is less desirable, bitter winter or the oppressive summer.

This is the summer. The vermin are viciously active. Water is in short supply. The stench coming from a variety of directions is strong and nauseating. Now one more problem is added. Someone is sick. Medical attention is out of reach. No one knows what is wrong with the "old woman." She lies on her infested dirty straw, unable to move. She is too weak to defend herself against the crawling, invading pests. They take advantage of her. They are merciless. Her face is red. Her fever must be high. Her lips are parched. She does not ask for water. There is none. She does not ask for anything. Only her distorted features tell that she is in agony. Her eyes speak, not her lips. Her eyes are conscious and knowing. Those eyes see the approaching end.

Only from time to time, she makes an effort and moves her lips. And it is then that those around her hear again and again the same whispered words: *Oh my God, my body may bring disaster to you, what will you do with my body. How will you manage . . . ?* The patient died. At night, secretly and in stages, they buried her dismembered body in the garden.

Inevitably as these Poles continued to protect Jews they were faced with changing situations, situations they could not and had not anticipated. Some of these had to do with the number of fugitives.

Indeed, the majority of rescuers (78 percent) said that the number of those they had been protecting grew beyond their initial expectations. At times this expansion was imposed on them by outside circumstances and against their will.

To illustrate, first, at the suggestion of her stepson Thomasz Jurski, Stefa Krakowska and her husband accepted three fugitives. Eventually these people begged them to take in some of their relatives. Gradually their number grew to eleven. As Stefa explained, an additional three came in a way that verged on blackmail:

I knew that this family of three: husband, wife, and child, were staying with a Polish woman who, according to rumor, was the Jewish man's mistress. She probably did not love him enough. . . . At any rate, they knew that I was hiding Jews. This, of course, was bad. . . . One day I looked out the window and I saw this woman come to the front of my house with the Jewish family. As they reached my house I saw the Polish woman nod and walk away. These three Jews, the Ziemans, stood there. They simply stood there looking at my windows in front of my house. What could I do? I had to let them in. If someone had discovered them there and they talked we could have all died. You see, they knew about the Jews I had, and when their Polish woman did not want to keep them, they asked her to bring them to my place. This was blackmail, and I could do nothing. I took them in. As it turned out, they were terrible people. . . . They continued to blackmail me, demanded special food, and threatened that they would scream or leave, which would endanger the rest of us. In fact, with the first Jews that I accepted, each additional person they wanted me to take was in a way a blackmail.. . . To be sure I also felt sorry for them, very sorry. But when I took in the first Jew I signed a death sentence . . . from that moment I was in the same boat as they.

Stefa's case was unusual. The differences between her charges and the manner in which they came created special tensions. Not only did some of them make unreasonable demands; they also quarreled continuously among themselves. When their fights reached an unbearable level, Stefa and her husband put a stop to it by threatening to move out of the house. If carried out, this threat would have left them without protection.

Tomasz Jurski, Stefa's stepson, reconfirmed the story: *It so happened that these people quarreled among themselves. At times they were almost losing their minds. It was hard. Some of them had money, some did not. Some ate well, others did not. Finally one of them instituted a commune, all had to eat the same and share . . . .* Tomasz told me about another serious complication. Twenty years older than his vivacious and outgoing wife, Tomasz's father was a retiring, quiet man, serious and very different from the fun-loving Stefa. One of the fourteen fugitives was a young man, Zenek. It soon became an open secret that Stefa and Zenek were lovers. Tomasz thinks that his father was aware of the affair but was too proud to say anything and too decent to act upon it. The other charges, however, became alarmed by this liaison and pleaded with Zenek to stop it, but to no avail. The love affair, and with it a tense and potentially explosive situation, continued. Only the end of the war finished the illicit romance; at that time, too, Stefa's marriage came to an end.

Janka Polanska, another courageous helper, also had her share of disappointments. Apart from the constant bickering, other actions of her charges shook Janka's trust in humanity. One day she brought home a distraught old Jew whom she had found on the street. Janka was appalled by the hostility with which those she sheltered greeted this unfortunate individual. All ten objected vigorously and openly to the man's presence on grounds that he increased the danger of discovery. To avoid complications, Polanska succeeded in finding him another hiding place.

Rosa, one of the ten protected by Janka, was a young woman who looked very Polish. Unlike the rest who had to hide, Rosa moved around freely, pretending to be Janka's maid. Because all of them had some money, it was decided that each should contribute equally to their living expenses. Janka and the make-believe maid would do the food shopping. One day Janka was surprised and hurt by Rosa's suggestion that they charge the others more money for the food and pocket the profits. To Janka's indignant refusal, Rosa only gave a disappointed shrug.

At times, too, the rescuer–rescued relationships underwent unusual changes that defy simple classification into a positive or negative experience. This happened to Eva Anielska, who was married to a young Pole who wholeheartedly approved of her extensive help. Among those the couple kept in their apartment was one Jewish man with whom Eva fell madly in love. When the war ended she divorced the Pole and married the man she had saved. They had two children.

However, his newly recaptured freedom went to her husband's head. He insisted on living to the fullest, and for him this meant having as many love affairs as possible. Eva found this intolerable. A divorce followed. Eva never mentioned this part of her life to me. I only found out about it by chance from archival materials at Yad Vashem in Israel.

Among people who live in close proximity and under conditions of continuous threat, strain becomes inevitable. Still, the rescuer-rescued relationships were special in that invariably pain and disappointment combined with positive experiences, attachments, and caring. Thus, for example, Stefa Krakowska moved to Israel at the invitation of some of her charges. There, while continuing to complain about them, she also continued her close association with them. Stefa's relationship to these people resembles family attachments. She cherished them but she was also disappointed by their shortcomings. No doubt, she demanded of them more than she did from most other relationships.

Similarly, Tomasz, Stefa's stepson, after he had finished telling of the difficulties, added: *Except for the love affair, the relations between my family and most of those hidden in our house were close and warm. They usually bickered among themselves.*

As the result of a routine raid, Tomasz was sent to the concentration camp Auschwitz. After his release at the war's end he went to meet one of the Jews they had protected, now a prosperous factory owner. This is how Tomasz described their reunion: *Imagine, someone throws himself at my neck and screams: "You are alive!" I was so moved I cried. All were looking on, watching their director make a fuss over someone who looked so poorly. . . . He took me home and his family was happy to see me. They wanted to keep me, but I did not want to stay. They were very good to me. I had dysentery—they nursed me.*

Janka Polanska also insisted that after the war she felt most relaxed in the company of those she saved. She felt that, as a group, those she had saved became her second family. In her case an estrangement from the rest of Poles followed.

When questioned about the quality of relationships to their charges, rescuers were reluctant to complain. Direct questions about problems elicited denials. I learned about such strains and conflicts only indirectly. These negative features, however, never dominated the relationship. On the contrary, the predominant views about those they saved were positive. In fact, only a small minority of the helpers I interviewed expressed resentment toward those they saved.[8]

Similarly, nearly all of these Poles were content that they had chosen to protect Jewish lives. Only seven of them had misgivings; these few felt that perhaps they had endangered their families too much, or that they should have taken more precautions. None of them regretted having saved lives.[9]

No doubt these Poles knew, just as most people do, of the inevitability of conflicts in close relationships under adverse conditions. Why, then, were they so eager to deny the presence of conflict? Why did they object, even resent, my questions about it?

It is unlikely that their denial of difficulties was prompted by a desire to distort reality. Rather, these rescuers reacted to my questions with the same resentment they would feel if a stranger were to try to dissect their family relationships.

Moreover, for these Poles the rescuing of Jews was a highly valued activity. To dwell on the shortcomings might have detracted from the emotional investment and the many sacrifices that were a significant part of this relationship.

Finally, my impression is that the rescuers' tendency to focus on the positive, rather than on the negative, reflects a realistic appraisal of the situation. Under the circumstances, to begin and to continue the protection of Jews, the rescuers must have had a preponderance of positive attitudes and feelings for what they were doing. The overwhelming majority of survivors (84 percent) reported that these Poles extended help willingly. Similarly, an overwhelming majority of the survivors (96 percent) felt that they had received "good" or "excellent" treatment. On balance, then, by focusing on the positive aspects of this relationship the rescuers were giving an accurate picture.

What happened when these predominantly positive actions and reactions were put to the test by special outside dangers? Did the rescuers waver in their determination to save?

Death was constantly hovering over the rescuer-rescued relationships. To avoid it Jews had to be on the run. All evidence shows a continuous pattern of changes and movement. Of the survivors, only a small minority stayed with one helper; the rest were forced to find different accommodations. Some had to move as many as twenty-five times.[10] Still, despite the overpowering odds against them, most Poles tried to continue their aid.

At times even when faced with grave dangers some helpers chose to continue their protection.[11] The reactions of Jan Rybak, the peasant who saved Pola Stein, illustrates this choice.

One winter night the Nazis moved into the village looking for Russian partisans. As they searched one house and then another, news about their raid spread throughout the village. The Steins and their protector knew that it was only a question of time before the Nazis would pay them a visit as well. They also knew that the attic, which up until now had served as a hiding place and protected Pola and her father from inquisitive neighbors, was unsafe. It was unrealistic even to hope that the Nazis would not discover the place. Therefore, together, they decided that Pola and her father should move to the nearby forest, and return after the raid. Father and daughter dressed warmly against the winter night and were ready to go. Outside a blizzard raged, making furious and terrifying noises.

Eight years old, Pola was afraid of the dark, the cold, and the forest. She was aware of what was happening and clung to her father's coat for protection. She tried to act bravely, but when the time came to hug the Rybaks goodbye, hot tears made their way down her cheeks, slowly and silently. Through misty eyes she noticed with surprise that Jan, their strong and independent host, was also crying, openly and without embarrassment. Then, almost in defiance, he reached for her hand, then to his wife he said: *They were so long with us, if we have to die we will die together. The child will freeze in this snow!* They stayed and the Nazis never came to their farm. Most other accounts also show that the decision to continue their protection in the face of special dangers was motivated by compassion and courage. This decision could be taken even when their charges urged them to part company. The rescuer Emil Jablonski felt close to his friend even before he came to stay with him. His protection only deepened their friendship. Despite all the threats and dangers he never thought of turning his friend out.

In 1944, after the Nazis crushed the Polish Warsaw uprising, they began to evacuate the civilian population to concentration camps. To avoid this dismal fate the Jablonskis decided to escape from Warsaw and into the part of Poland occupied by the Russians. They were aware that Kazik's typically Semitic features would become a special liability. In fact, because of this all of them could die. To shield his friends from disaster Kazik begged them to leave him behind, hidden in the cellar.

But Emil would not hear of it: *I insisted that he come with us. He was very impractical and helpless. After all, it would have meant death for him if he stayed alone in the cellar. I knew too that because of his Semitic features it was risky to take him with us. I could not leave him. . . . In the end the four of us, dirty with collars up, kept moving from place to place among the ruins of the city. . . . We managed.*

The Jablonskis and Kazik succeeded in eluding the Nazis. Away from the capital Kazik, who suffered from tuberculosis, became seriously ill. Emil, with great difficulty, managed to hospitalize him. But all Emil's care and attention could not save his friend. Kazik died a few weeks before the war ended.

Depending on state of mind, we tend to relate and experience time differently. When we are content and happy time seems to fly fast. In contrast, when faced with grave, life-threatening dangers a minute may seem like eternity.

In those uncertain days, time lost some of its conventional meanings. Especially for the rescuer-rescued relationship, time could be at once the enemy and the redeemer. Filled with awesome possibilities of sudden change, time seemed to refuse to move, stretching every minute into an eternity, becoming both static and dynamic. A helping act, no matter how brief, could make a difference between life and death. While a longer period of protection led to a sense of security.

Apart from the way they felt about time, on balance for rescuers long-lasting aid had to be more perilous. Simply put, more things could happen in a longer period of time.

Did the inherent danger of long-term help result in a preponderance of short-term acts? Some have assumed this to be the case.[12] But my evidence tells a different story. The majority of rescuers I studied extended their aid for more than six months, while a very small minority limited their aid to single acts. Jewish sources confirm these results: among the survivors single helping acts were least common.[13]

Contrary to what some had expected, the rescuers defied time just as time tends to defy our own wishes: fleeting when we are happy and wish it to linger, refusing to move when we are threatened and deprived. Throughout their prolonged rescue, Polish helpers refused to be intimidated by time. In the end they succeeded in defying both the dangers of rescue and the capriciousness of time.

Just how time could make a difference between life and death is illustrated by Franek Dworski's experience. During the war Franek, his wife, and little daughter shared their modest house with many who had to hide from the Nazis. Most frequently their illegal guests were Jewish; at times they were Polish underground fighters. For those who turned to him, Franek performed all kinds of services. To some he offered food and shelter, for others he arranged false documents, for some he found employment, some he would escort from place to place. There was hardly a task Franek would have considered too difficult or unworthy of his attention.

On a particular fall evening, about half an hour before curfew, a strong wind, mixed with an equally strong rain, created a strange and threatening backdrop. In sharp contrast to the outdoor fury, the Dworskis' household was peaceful. At that point only two Jews who were passing as Christians shared their home. Both had false papers and both were officially employed, which meant that during a raid they did not have to hide.

On that evening as all of them were gathered around the table for their customary dinner of potatoes and dark bread, they heard a hesitant tap at the window. The same tap was repeated at the door. No one was either surprised or frightened, because people often came there for help. The tap itself was too uncertain, too gentle, too apologetic to suggest that it was anyone who intended to harm them. In any case, those present had relatively little to fear.

Indeed, the opened door admitted three harmless-looking beings: a man, a woman, and a girl not more than seven. The trio hesitated before entering and once inside stood, uncertain, huddled together, while the water from their soaked-through clothes formed puddles on the floor around them. They appeared exhausted, sad, and resigned. The child's red cheeks, shiny eyes, and unfocused look betrayed a high fever. At first speechless, they were soon encouraged by Franek's warm smile and welcoming greetings. They had been directed here by one of Franek's friends and explained that they would like to rest for one night only because a place was waiting for them in the country. Since they had no documents, they felt that it would be safer to reach their new shelter in daylight. They added that they had been forced to leave their present home in a hurry because they had been denounced.

Without hesitation the Dworskis agreed to let them stay. Right away their clothes were set out to dry. They were fed and told that they could sleep in the attic. Franek also explained that since their house was safe, there was no point in the three rushing off. On the contrary, they could rest here for a few days, at least until the child felt better. After that Franek would personally bring them to their new place. Grateful, the strangers went to bed.

At dawn violent knocking shook the entire house. Then heavy boots began to kick the door. Holding on to her daughter, Mrs. Dworski admitted five Gestapo. They pushed her roughly aside and spread swiftly around the ground floor. As they began to search they asked: "Where is the Jewish family? Where are you hiding them?" Clearly someone had tipped them off.

Then, without glancing at those present, the Nazis climbed the ladder to the attic. Left behind, the rest remained immobile and silent, avoiding each other's eyes.

Franek Dworski recalled: *Here were my wife and daughter, clinging to each other, both pale as if the blood had decided to leave their faces. I felt sorry for them—actually I personally had no regrets, only a wave of sadness for my wife and child came over me. This was it. Then I felt nothing. I still felt nothing when, obviously angry and disappointed, the Gestapo came down from the attic.*

What had happened? The Jewish family had sneaked out earlier, leaving the place in perfect order. The Dworskis later found an unsigned note under the mattress thanking them for the warm hospitality and explaining that they were leaving because they did not want to endanger the entire household.

Had the Dworskis refused to accept this Jewish family, these fugitives might have died. On the other hand, had this family stayed a few more hours this could have led to the death of eight people. Much could have happened in a brief span of time.

Both descriptions of the specific kinds of aid and their consequences could be greatly multiplied. From the rescuers' perspective, all protection of Jews was hazardous. How many Poles had actually faced these life-threatening situations?

How many righteous Christians were there? Exact figures are elusive. What we have instead are estimates that vary with the particular assumptions on which they are based.

By the end of 1984 the Yad Vashem committee had bestowed the title of "righteous Christians" on 5,742 individuals. More cases are pending before the committee. The requirement that the rescued Jew and not the rescuer apply for this title deprives many deserving people of ever receiving it. Some Jewish survivors are dead, others have lost touch with their protectors, others simply refuse to apply, and still others are not even aware that such a distinction exists.

As for the Christians themselves, not only have many of them died since the war, but some lost their lives during the war precisely because they rescued Jews. Many such helpers were never recognized by name and never will be. Dr. Paldiel, the director of the Department for the Righteous at Yad Vashem, estimated that to obtain the real number of righteous Christians one would have to multiply the number of those who received this official designation by ten.

Of the 5,742 who were already honored by Yad Vashem, 1,505 come from Poland.[14] Some have argued that since each Jew who tried to pass must have benefited from the aid of a few individuals, the number of Polish rescuers runs into hundreds of thousands. It is, however, possible to argue differently. Since at times a single Pole extended help to large numbers of Jews, there may well have been fewer Christian helpers than passing Jews. Evidence supporting each argument is available, with few hard facts to substantiate either assertion.[15] Still, there is no disputing that those who have been recognized by Yad Vashem represent but a fraction of those who deserve to be. Of the 189 rescuers in my study, all of whom fit the definition of righteous, less than a quarter have this official title. Of the survivors, very few refer to the Yad Vashem distinction; less than 20 percent. Of those who do, an equally small percent say that their protectors received this honor.

Earlier I described righteous Poles who lost their lives because they were protecting Jews. As a rule they died together with their charges, and Yad Vashem distinctions thus will not be offered to them. How many such Poles were there? My efforts to count some of those who died while saving Jews and who were identified by name have resulted in numbers ranging from 343 to 668.[16] Recalling that at times the Germans would surround and burn entire villages accused of harboring Jews, annihilating every inhabitant, it is obvious that neither the names nor the numbers of the resulting fatalities can ever be known.[17] For these and no doubt other reasons, only few of the Polish righteous can be recognized and honored.

I believe that knowing the exact number of the righteous is not as important as understanding who they were and what motivated them toward this life-threatening behavior. Thus, rather than continuing a futile search for their elusive number, I will explore and examine their characteristics and motivations.

PART II

The Exceptions:
Paid and Anti-Semitic
Helpers

CHAPTER 5

===

The Issue of Money

Why should I be grateful? . . . he loved money . . . he did it only for money; besides, every week he kept raising the price. . . . He used to tell me that if the war would drag on he would not keep me. Lola Freud did not approve of the man who gave her shelter.

Roma Zelig's experiences differed; the peasant who saved her never raised the price. And even though Roma describes him as a decent man, her reaction is similar to Lola's: *He did it for gold, not because he liked us; then he became scared and fed up. If the war had lasted longer he would not have kept us. . . . This is why I really don't feel gratitude or whatever. It was strictly a business deal.*

Annoyed by my questions, Roma told me how she and her family of seven lived in constant fear: *Quite often he would say that he did not know why he was sticking his neck out for Jews. . . . He took us for a month, but he never meant to keep us that long . . . he was trapped . . . there were times when we thought he was planning to be rid of us. We were a burden. But it was not easy to do away with eight people.*

Reactions such as Roma's and Lola's are very common. They illustrate that help motivated by money left little room for gratitude and attachments, and that resentment grows when people's lives are reduced to a business proposition.

A similar negative evaluation of this kind of aid is also suggested by the Yad Vashem Commission, which bestows honors on Christians who saved Jews. As I mentioned earlier, this commission automatically denies recognition to those who protected for personal financial gain. In my study I, too, excluded from the definition of rescuers all who saved because of payment.

The need that a sharp distinction be drawn between those who protected for gain and those who were motivated by very different considerations comes yet from another quarter. Righteous Christians are also eager to keep the two types of aid distinct. After the war, many who had selflessly saved were subjected to repeated innuendos by their fellow countrymen that such aid had really been given as part of a secret exchange of monies. Visibly upset by all such attempts, no matter how subtle, to so link their aid to Jews, these rescuers described their frustration in trying to persuade others that they had done what they had done for no financial gain.

In Warsaw, Hela Horska told me how after the war other Poles refused to believe her explanations for aiding Jews: *For many years I was accused of having saved Jews for money. It was like a curse hanging over me. There was no way in which I could have made them change their mind . . . they simply could not get it into their heads that others would risk their lives out of the goodness of their hearts.*

In New York one of Hela's charges, David Rodman, told me how she had, in fact, kept her protection of Jews apart from financial considerations. David remembered hiding in a room when a Jewish man came to Dr. and Mrs. Horski to ask for shelter: *This man said, "I know you are helping Jews. I will give you $25,000 if you will help me and my wife!" I knew who it was. She helped that man's two brothers later on. She said: "Sir, do you think that for $100,000 I would lend you my children? If I do help Jews, I want to, and if I can I will do it without money. It is almost insulting, because my children are too dear to me. You know that I can lose them. So it is not a question of money here. It is a question of how much I can do, to whom. I have preferences. I just cannot take everybody."*

Many other righteous Christians demonstrated, as Hela Horska had, that money played no significant role in whom they helped or in how generous they were in extending aid and protection. For example, Ada Celka and her sister, poor and single, were employed as governesses. They were also taking care of a paralyzed father, with whom they shared a one-room apartment in a poor section of Warsaw. During the war they added one more lodger: a young Jewish girl, Danuta Brill.

After I interviewed her in Warsaw, Ada gave me the address of Danuta Brill. Only upon meeting Danuta in New York did I learn that the Celka sisters were asked to take in a Jewish boy whose parents were willing to pay them a lot of money. Since they could not safely accommodate one more child, they refused the offer and continued to keep the penniless Danuta.

During the war, single and poor, Janina Morawska lived close to the forest in a one-room hut. At night she was frequently visited by hungry and homeless Jews, whom she offered food, and occasionally shelter, especially on cold winter nights. For nine months, Janina also kept in her hut a Jewish woman, a virtual stranger left there by her sister. Regarding money she said:

I never could have taken a penny from these unfortunate people. Although I was poor I was not greedy. My conscience would not have allowed me to take money. I was offered by different Jews gold and jewelry, but I always refused. I simply could not take any of it. "Take it with you," I told them, "you may need it to save your life." Most of these people lived in horrible conditions in the forest—they were so vulnerable, so exposed to danger. Then, almost as an afterthought, she added, *it is horrible to want to make profit out of someone's misfortune.*

Indeed, for those who were fortunate enough to come across the righteous, money might have been less essential to survival than other factors. The reaction of one rescuer, to be sure the only one, suggests that wealth could become a disadvantage. Eva Anielska noted that during her activities in the Warsaw ghetto, *rich people would approach me with offers of money. I refused to help them. . . . No one would help the poor, so I had to. The rich could get contacts for payment.*

Implied in these consistently negative reactions to paid rescue is the idea that among people for whom personal integrity is important, money can play only a very limited role in determining moral conduct. For such people to have gained financially from the helpless and dependent would have been morally repugnant. Such behavior would also have been incompatible with the meaning of rescue. As important, by refusing to attach a financial price to their self-sacrificing behavior, these Poles reaffirmed by their actions the view that the value of life is on a different scale than that of money. At the same time they demonstrate that people may and do sacrifice their lives for principles, for ideals, and for other human beings, but may be less likely to do so for payment.

Yet the fact that there were some helpers who rescued selflessly does not change the fact that there were others for whom money was the much more important, if not paramount, reason for helping. On the contrary, the widespread impression among many who have studied the Holocaust has been that because there were some who were motivated by money, there were hardly any whose motivation was unselfish.[1] The literature shows examples of both: Poles who saved Jews only for money, and Poles who courageously provided Jews with protection, their actions unrelated to economic considerations and at great risk to themselves and their families. But this literature, which consists of random accounts from individual case histories and memoirs, cannot tell us much about the extent and implications of paid assistance as opposed to unpaid rescue. For that, a more systematic inquiry is required.

In talking about paid helpers, it is important to establish that I have applied the term only to those individuals whose aid to Jews was motivated solely by financial rewards—that is, only those who would not have given aid in the first place had there been no payment. Not included in this category are cases where payment was made but where such payment was not the main reason for Jewish protection. If, for example, a Jew had money and a Pole did not, it was natural for the former to pay. This payment sometimes covered the expenses of the fugitive; in other cases, it might have supported the rescuer as well, but payment in itself was not the reason for helping. In short, all such cases are still defined as righteous.

Next, it is also important to note that information about paid helpers is not as complete as information about the righteous. None of the Polish rescuers in my study, for example, protected Jews for money.[2] Thus their experiences are not applicable here. Nor is there extensive archival material on paid rescue. In general, those who saved for money have not identified themselves and those few who have been identified by others have refused to discuss their experiences. None wrote wartime memoirs, for example, describing the nature of their aid to Jews; nor did any testify before the many postwar historical commissions that collected this kind of information. Perhaps the paid helpers' refusal to discuss their role suggests that they too share negative perceptions about rescuing for money.

But although paid helpers may have reasons for refusing to discuss their protection of Jews, paying Jewish survivors have no reason to remain silent about their experiences. Therefore, Jewish accounts become an indirect source of information about paid helpers. And because most passing Jews experienced help from more than one person or family, many of those who might shed some light about paid help do so in the context of having also received unpaid aid from righteous people.

Using as a source of information the group of survivors in my own study, I find that they report that only 16 percent of those Poles who agreed to help them conditioned their help upon payment (i.e., were paid helpers). Of the remaining 84 percent, over half actually assumed some financial responsibilities for their Jewish charges and almost a quarter accepted no payment at all, but it is hard to establish whether this quarter supported the fugitives or not. Finally, somewhat fewer than 10 percent took money from their charges, but payments were not the reason for protecting Jews.[3]

How did the contact between Poles and paying Jews begin? Just as with rescuers, Jews initiated the relationship with paid helpers by approaching the Christians and offering to pay them in exchange for aid.[4] Poles who saved Jews for payment were more likely to pledge this aid beforehand; however, they were also more likely to withdraw their offers. Some robbed the Jews of their money before reneging. Considering how many more paid helpers than righteous Poles went back on their promises, perhaps some of them had no intentions of honoring their pledges in the first place.[5]

Who among the Poles was most likely to be receptive to such an overture? Since no underground organization charged for rescue, paid helpers all fall into the category of individual helpers. As a group, these individual helpers appear to fall into the socioeconomic category of "poor and very poor." More of the paid helpers were poorly educated, and not unexpectedly, their occupations are less prestigious than those of the righteous. The majority of paid helpers were peasants while only a minority of the other helpers were.[6]

Prompted by different motivations, did paid helpers and righteous Poles also differ in terms of the kind of aid they offered? At one level the answer is no. As with the majority of rescuers, few of these paid helpers limited their protection to a single act, and in terms of duration, over one-third shielded Jews for six months or longer.[7] The survivors' reports show that sheltering—one of the most important forms of rescue—was in fact more common among the paid helpers than among the rescuers.[8] In addition, paid helpers often escorted their charges to safety, and many of them warned Jews about impending danger.[9] However, only 11 percent are said to have offered food or money.

But if, from afar, paid rescue and unpaid rescue look somewhat similar, up close they were very different. Exactly what kind of relationship existed when payment defined the nature of the bond? Did proximity of Jew to Pole and the Nazi threat to both lead in paid rescue to the mutual care and attachments that existed in unpaid rescue? A look at the experiences of those who were protected for payment may suggest some answers.

A teenager at the time, Ida Brot, along with her brother and mother, were hidden by paid helpers. When I asked how she was treated, she flared up in anger: *How can you even ask such a thing? Don't you know yourself? Surely they did not keep you for love.* Even more sarcastically she added: *Do you want to tell me that the Poles were good to you?* She became visibly more irritated when I told her that I was treated decently.

Ida then told me how soon after their arrival the first peasant demanded that they hand over all their jewelry and money. When he received nearly all of it (some they had managed to hide), he proceeded to search through their belongings: *When he had convinced himself that he had taken all our valuables, he asked us to leave. He insisted that it was too dangerous and that we had to go. In fact, while we were staying in his place he told us about a Jewish doctor and his family who were being hidden by a local peasant. Once this peasant had stripped the doctor's family of all their possessions he threw them out. The doctor and his family went to a nearby forest. After they left his house, the peasant denounced them. As a result, the doctor, his wife, and their three children were shot. Our peasant felt that this incident had increased the danger and that we had to leave his place.*

When, in desperation, my mother asked him where we should go, he shrugged his shoulders and said: "That is not my responsibility." We still stayed on for a while, but when he became insistent and began to threaten us, we had to leave. We were afraid that he might kill us. One night we simply slipped out of the barn.

All the time we stayed with this peasant we were treated poorly. The food he gave us was inferior and meager. He put us in a barn where we became dirty and infested with lice. Once I became very ill, with a high fever. The peasant showed no compassion at all. Only after much pleading and begging did he agree to bring one aspirin.

Sonia Lieber's story, although different, resembles Ida's. During the 1943 uprising in the death camp Treblinka, Sonia managed to run away with two others. As they were roaming the countryside a Pole advised them that they would be safer in a nearby forest. In the forest they succeeded in establishing contact with a Polish peasant who was willing to give them food in exchange for gold. Winter was approaching, so they asked the woman for shelter. Sonia explained: *This peasant, a primitive and cruel woman, had in her place three other Jews, whom she kept in terrible condition. She had agreed to let these Jews stay with her only because they promised to give her a house and other property after the war. In our case, too, she agreed to take us in on condition that we pay her well.*

But two or three days after she took our gold and a diamond, she started to complain that what we had given her was worthless. We made the mistake of giving it all to her at once. . . . Actually we had no choice, she told us to. Besides, one of the major items was a big diamond, which could not be subdivided.

In time she became more horrible, giving us practically no food and complaining ceaselessly about the sacrifices she was making for us. In fact, one of her favorite expressions was: "I am feeding my own death."

In January she insisted that we leave, telling us that the Germans were about to make a raid and that our presence would endanger her life. Actually at that point, the Germans were organizing a hunt. She was no doubt hoping that they would hunt us down like animals. At that point she also threw out the Jews who were there before us. One of them, an old and sick man, instead of moving with us decided to stay behind in the forest. He refused to come and we never saw him again. The rest of us after a few days, hungry, half-frozen, crawling rather than walking, returned to the peasant's place. We moved back to the attic in the barn.

It seems that she was not quite ready to kill us. And so, begrudgingly, she kept us on. But she tried to starve us out. Each of us was allotted three potatoes a day and water. And as she did deprive us, she never tired of repeating what a terrible burden we were to her. There was no pity at all in her, no mercy. Finally, in June, we decided to go into the woods again because we felt that we were being killed by starvation.

When we did leave, we could barely walk. We were weak and sick. At first there were six of us; only three survived. It was a miracle that we survived at all, inexperienced as we were with life in the forest. We fed on eggs from birds' nests and on berries.

That summer Sonia Lieber and her two companions were liberated by the Russians.

The experiences of a woman named Rosa Pinczewska with a peasant who protected her and her family in exchange for exorbitant prices point to how easily paid aid could result in a disastrous turn of events. One day Rosa's father, who usually went to the village for food, failed to return. Shortly after that, their financial resources dwindled to nothing. It was at that time that the peasant on whose farm they stayed put Rosa, her mother, and a younger brother into a wagon, explaining that he was taking them to a safer place. When they reached a river, the peasant forced them out of the wagon and then proceeded to drown Rosa's mother and brother. Miraculously, Rosa escaped. She reached a village and there, without revealing her identity, was hired as a farmhand.[10]

If I move from the individual case to the overall figures I find that among the survivors who were protected by paid helpers, 59 percent believed that these Poles had negative attitudes toward them. In contrast, none of those who benefited from the protection of the righteous felt this way. Negative attitudes translated into negative actions. Of those who had paid helpers, 64 percent said that they were treated poorly, while only 3 percent of those who were saved by rescuers felt this way. Mistreatment could be expressed in a variety of ways: starvation, robbery, and demands for higher than agreed upon pay. A substantial proportion of these fugitives experienced all these abuses at once, and only a minority managed to escape such mistreatment.[11]

Not surprisingly, Jewish survivors also resented their paid helpers. Most of them defined their paid helpers either in overall negative terms or as "greedy" and "money hungry." Only very few saw them as "courageous and good-natured." Yet nearly all righteous Poles are described in the most flattering terms.[12]

In many regards, these descriptions are accurate. Only those who aided Jews for money kept increasing the price and only they threatened their charges repeatedly with denunciations, threats which we know some followed up on.[13] But what about those paid helpers who tried at least to some extent to live up to their end of the arrangement? Where do these paid helpers lie in comparison with those who at one extreme took advantage of their helpless Jewish charges practically from the beginning, and with those at the other extreme, who were righteous Christians?

As I mentioned earlier, most of those who aided for payment were very poor and largely without prospects. Initially these Poles saw in the paying Jews their only chance of escaping from an unbearable situation. For some of them, economic improvement meant realization of a distant dream. However, once the dream was fulfilled, once they became better off, a different kind of reality began to reassert itself. It was a threateningly awesome reality that could not be ignored.

With economic improvement, which had begun with Jewish presence, the very reason for this presence was gone. And once the desire for payment was satisfied, there was nothing to neutralize the existing threats and dangers. All that was left were the life-threatening forces. How did paid helpers react?

As I have already pointed out, the Nazi law specified that those who aided Jews in any way and for any length of time were committing a crime punishable by death. Perhaps initially Poles who agreed to harbor Jews for payment simply refused to focus on the Nazi prohibition. Some may have felt confident that if trouble came they could handle it. Others may simply not have anticipated the many ramifications of this very important decision. In time, however, the existing threats and dangers multiplied and usually at some point could no longer be ignored. At this point for the Jewish fugitive the difference between paid rescue and unpaid rescue often meant a difference between life and death.

Theoretically, once a paid helper wanted to rid himself of his Jewish charges, he or she could do one of three things. As we learn from the case of Rosa, one way for a paid helper to rid himself of his charges permanently was to murder them. Just how many Jews were so murdered we will probably never know, but we do know that at least some were. Other than that, a paid helper had two other choices—he could throw out his charges, telling them to fend for themselves, or he could denounce his charges to the Germans.[14] But to do either of the latter was not without risk Poles could be executed for having extended help in the past.

It is possible that some Poles chose to murder their victims rather than send them away or denounce them because they feared that, if captured, these Jews might reveal who had protected them.

But even paid helpers who were ready to murder their charges were sometimes confronted by unexpected hurdles. This is illustrated by the case of two Jewish men who were hidden in a cellar by a Polish janitor. With time this paid helper became progressively more apprehensive until he decided that the safest way out of his predicament would be to murder the fugitives. When the two men were asleep he sneaked into their hiding place, and with the swift blow of an ax split the head of one of them. Faced with the gruesome picture of an open head, overflowing with blood, and the shocked eyes of the other prospective victim, the murderer began to tremble, unable to move. He then heard himself say: *I cannot kill anymore but if I don't kill you and if you should survive you will give me away.* When the fugitive promised never to reveal what he had witnessed, the janitor agreed to spare his life. This paid helper continued to care for his charge till the end of the war.[15]

To be sure, a Pole denounced by a Jew could deny the charges; yet there was no way to know whom the Nazis would believe. Aware of this possibility some Poles reacted very differently—by continuing to hide their charges despite a desire to be rid of them. Indeed, those who were incapable of personally murdering their Jewish charges but were yet unwilling to face the risk that they would be murdered or punished by the Germans should their past protection be revealed, continued the protection but felt trapped. And so, within this relationship of little freedom, what few options there were involved considerable risks. Regardless of how the Jews and Poles felt about each other, their ties, once formed, could not be easily cut.

The case of Oskar Pinkus and his family of seven illustrates this situation well. The Pinkuses had paid an impoverished peasant to hide them. For this man and his family the arrival of paying Jews averted starvation. Once, however, the well-being of these paid helpers became a reality, they listened with trepidation to stories about others who had lost their lives while hiding Jews. Moreover, from the perspective of the head of this peasant family, the end of the war was nowhere in sight. With diminished economic worries, anxiety about personal safety and survival began to take precedence. When the peasant eventually asked the Pinkus family to leave, they ignored their host's request and continued to stay, having no place to go. Not about to give up, the peasant decided to scare them with tales about impending raids. When these measures failed, he reduced their already meager food allotments. In the end he became abusive and began to threaten them.

At this point the Jews asked: *If we leave and the Germans catch us, don't you think they will force us to reveal where we were hiding all this time? The farmer was dumbfounded . . . he sat there confused and frightened not knowing what to say.*[16]

Oskar and his family stayed until the end of the war. Their reactions to the Pole's demands, however, were not without risks. After all, the paid helper could have stopped feeding them altogether. Or he could have killed them.

Besides, subsequent events only emphasize the complexity of the Pinkus situation. In 1944, a Polish underground unit was in control of the area. Even though this peasant knew that these underground fighters were killing Jews as well as Germans, the peasant made a special effort to protect the Pinkus family rather than turn them in.

Paradoxically, the same Pole showed little compassion to these unwanted tenants. While raising his prices and starving them, he voiced openly his regrets at having become involved with Jews.

Knowing that some Jewish fugitives refused when urged by their hosts to leave, I asked Ida Brot why, when the first peasant insisted that she and her family leave, her mother did not hint at the possibility that if caught they might be forced to denounce him. Ida was shocked by my question: *Oh no, he would have killed us if we said anything like that . . . besides, my mother was not this type of woman.*

The experiences of the Pinkus and Brot families and of Sonia Lieber were not at all unusual. Negative attitudes and mistreatments by paid helpers were often expressed by the desire to be rid of their charges. Moreover, many Jews who paid their helpers (19 percent) considered the special dangers cited by Poles as imaginary. In contrast only three individuals who were saved by rescuers, or 1 percent, expressed the same feeling. Objectively, of course, there is no reason to assume that paid helpers were in fact subjected to more dangers than righteous Christians. Yet the fact that most paid helpers wanted to disengage themselves from their obligations to their charges because of real or imagined dangers, while most righteous Christians did not, underlies most clearly their different attitudes to, and evaluation of, Jewish rescue.[17]

Not all paid helpers who asked their charges to leave did so out of resentment or disapproval. In fact, the desire to break the relationship did not by any means always develop in an environment of mistreatment by the paid helper. It is nevertheless true that no matter what the special reasons, helpers who were paid showed greater eagerness to discontinue their aid. The experiences of Frida Nordau, who was protected by both paid helpers and righteous Poles, none of whom mistreated her, can shed some light on this distinction. The paid helpers asked her to leave, and after much wandering Frida found Poles who hid her free of charge. She emphasized that even with special raids in progress these unpaid rescuers insisted that she stay, arguing that they ought to face the difficulties together.

I believe that the paid helpers' failure to view the protection of Jews in other than commercial terms made them less able to withstand outside pressures. The preponderance of rational economic considerations rendered fragile the ties between paid helpers and their Jewish charges. Sometimes this fragility led to tragic consequences.

Sara Federer, the survivor I met in New York, was married to a physician from an assimilated family who could have easily passed for a Pole. After some experiences with blackmailers, however, he insisted on going into hiding. A Polish woman agreed to hide Dr. Federer and his friends, a married couple, in her cellar. Although based purely on commercial considerations, the relationship between this Polish woman and her charges was polite and free of conflict. However, at the beginning of the Polish uprising in Warsaw in August 1944, this paid helper ran away, without warning and without making any provisions for the three fugitives. They continued to stay in the cellar. Sara described what happened: *The Polish woman simply left him, his friend and wife, without telling them a thing. Then one day they heard spoken Russian through the window. My husband thought that the Russians had come to liberate them. Excited, he ran out thinking that the war was over and that he had made it. But these were the Wlasows [Wlasow was a Russian general who together with his unit defected to the Germans and collaborated with them]. A Wlasow soldier pulled out a gun and shot him right there in the head. When my husband's friends saw what had happened, they managed to save themselves. Later on, the man and his wife were evacuated to Germany as Poles. They tried to jump off the train and were both killed. A woman who was on the train with them survived. After the war she looked for me and told me how my husband died.*

Mietek Korn, passing as a Christian, was protected by his Polish looks and fluent Polish, but his wife had to hide because of her Semitic features. Through friends the couple located an impoverished Polish noblewoman who was willing to shelter Mrs. Korn for money. Later on their young son joined his mother.

All along this Polish woman treated her charges kindly and stuck to her initial agreement, never raising her price. Yet she was very outspoken about her anti-Semitism, emphasizing that she was saving them only because of dire economic circumstances. But neither the woman's anti-Semitism nor her outspokenness interfered with her proper and even kind treatment of the Korns. With the 1944 Warsaw Polish uprising, however, the situation changed. Mietek Korn, who did not stay with his family, explained: *The motive for keeping my wife and son was only money, quite a bit of money. . . . All the time they were there, there was no mistreatment. . . . When they (the Germans) started bombing this part of town their protector ran away and left my wife and son. My wife looked Jewish so the Polish woman was afraid. . . . My wife and son perished when the house they were left in was bombed. . . . I met the Polish woman later. She insisted that she could not have taken them; she was ashamed before her own family that she had tried to save Jews.*

But not all those who saved Jews for money mistreated them or left them behind when trouble came. In some cases warm and close feelings developed between Jews and their paid helpers with some Poles continuing their protection even after the paying Jew had run out of money.[18]

But such positive developments were rare. On balance, my findings consistently show that when money was the major motivating factor, the relationship between the rescuer and the rescued was less positive, less stable, and mutually less satisfactory. Moreover, because those who extended help solely for money do not fit the definition of righteous Christians, when possible I intend to use them as a "control group."

Why did the paid helpers feel and behave so differently? Answers are at once speculative and tempting.

Unlike life, money as a means to an end and its rational and quantitative nature precludes passion.[19] Free of emotions, money is an impersonal, objective commodity.

Life, on the other hand, is a most highly valued entity, and hence risking and sacrificing one's life involves a high level of emotionality. Because we tend to separate passion from rationality, we object when rational and emotional forces intrude upon each other. Given the rational nature of money, to use it as a basis for emotional actions would deny its essential quality. For inasmuch as life-threatening actions involve a high level of emotionality, money as a reason for such actions would undermine the very value of life.[20]

Paid helpers thus felt themselves caught in a vicious circle. With economic improvement, the reason for sheltering Jews disappeared, yet it was difficult to disengage themselves from this relationship. Such a disengagement itself involved tremendous risks and perils. In the absence of moral commitments, there was nothing to prevent these Poles from feeling anxious and trapped. Some reacted by asking the Jews to leave, some denounced, some murdered, and some mistreated their wards. For some, mistreatment might have become an outlet, a way of venting their pent-up frustrations and disappointments.

To be sure, the reactions of paid helpers were not homogeneous. Moreover, based on the survivors' accounts only, my evidence is limited. How many of the Jews who paid were betrayed or murdered by those whom they paid we can never know. On the other hand, in the few cases where a warm relationship developed, it served as a buffer against negative reactions.

The small number of Jews who were saved exclusively for money suffered because those willing to risk their lives for payment were soon caught in their own cycle of threats, frustrations, and helplessness. In this life-threatening and unstable situation, money could and did serve only as a transient and inadequate incentive, an incentive that soon denied its own importance.

Even in those few instances where paid helpers treated their charges reasonably well, the aid they offered in no way approached the commitment and self-sacrifice so clearly evident in the attitudes and behaviors of the righteous. Paid help was undeniably and consistently different from rescue. Hence the insistence of some that selfless protection of Jews was both very special and different from paid aid gains definite support. What other characteristics, besides their unique selfless commitment to help, do these righteous Christians share?

CHAPTER 6

The Rare Case of the Anti-Semitic Helper

As unexpected as it is intriguing, the rescuing of Jews by Polish anti-Semites has been a frequently recurring theme, both in the Holocaust literature and in the course of my interviews.[1] Interest in the topic, however, has not been matched by careful and systematic analysis. Basic questions remain unexplored: Who are these anti-Semitic helpers? How extensive was their participation in the rescuing of Jews? What does the rescuing of Jews by anti-Semites mean and what factors account for it?

Elsewhere I have argued that for most Poles it was difficult not to have internalized an anti-Semitic ideology. The very acceptance and pervasiveness of Polish anti-Semitism suggest that for the most part even those who were saving Jews might have in some ways shared in the prevailing anti-Jewish sentiments.* The inevitability of exposure to anti-Semitic influences raises the question of definition. Who is an anti-Semite? What criteria have been used when referring to rescuers as anti-Semitic? A closer look at the literature and my interview material suggest that those who are unaware of their anti-Jewish sentiments and values are excluded from this category. Excluded also are those who would not have supported or participated in any anti-Jewish actions. In contrast, anti-Semitic rescuers are individuals who through their actions or ideology had the reputation of anti-Semites, and who were aware of this image. In a sense, then, they are overt anti-Semites.

Who were these anti-Semitic rescuers and what exactly did they do? One of them in particular, Jan Mosdorf, appears to have caught the attention of all who touched on the subject. Mosdorf came from a socially prominent family. He was a devout Catholic, a national leader of the extreme right (ONR), and a distinguished lawyer.

Some survivors have interpreted Jan Mosdorf's help to Jews as proof that

---

*See my discussion in Chapter 3 of the diffuse cultural form of anti-Semitism from which even some rescuers could not escape.

righteous Poles could not be distinguished from others except by their very protection of Jews. One of those survivors, Aron Blum, referred to Mosdorf as an anti-Semite who repeatedly risked his life by sheltering Jews. When I pointed out that Mosdorf never kept Jews in his house, Aron accepted my correction, but countered that he protected Jews in other important ways.

Mosdorf's reputation as an anti-Semite was well known. He was also prominent politically. His strong patriotism and nationalism precluded co-operation with the Germans. Threatened by Mosdorf's political influence and independence, the Nazis sent him to Auschwitz. As a concentration camp inmate Mosdorf assumed an important position in the camp's underground organization. Soon he became known not only for his help to Poles but also to Jews. Unexpected as his aid to Jews was, it led to many exaggerated claims.

Felicia Zapolska, a rescuer I interviewed, was Jan Mosdorf's cousin. Felicia described him as noble, compassionate, and as someone who had sacrificed his life for others. She believed that as a leader of a rightist movement he would have been released from Auschwitz had he refrained from helping others, particularly the Jews.

Others who were at Auschwitz and observed Mosdorf's activities there paint a less glowing picture. Mosdorf, they said, did supply Jews with bread and carried letters for them, both highly valued activities. However, he performed these functions only sporadically. In fact, his priority was helping Poles; help to Jews came second. As an underground figure, one of his main responsibilities was to settle whatever conflicts arose between the Jewish and Polish inmates. At that stage Mosdorf became convinced that everybody ought to be united against the common enemy: the Germans. But just because he was advocating cooperation between Jews and Poles and because, among others, he was also helping Jews, did not mean that he renounced his anti-Semitism. One writer argued that he was only deferring dealing with Jewish issues for a more appropriate time.[2] For Mosdorf, however, the appropriate time never came. A Polish inmate denounced him as a Communist and Jewish sympathizer, and he was promptly executed.

Among the anti-Semitic Poles who stood up for Jews, the name Leon Nowodworski is also frequently mentioned. Before the war, as the dean of the Council of Lawyers, he supported the exclusion of Jews from the legal profession. Like Mosdorf, Nowodworski came from an upper-class family. He was also an influential and active member of the highly anti-Semitic National Democratic party (Endecja or SN).[3] Yet, during the war, despite his overt anti-Semitism, he openly disobeyed a Nazi order and refused to dismiss Jews from the bar, informing the Nazis that Poles would deal with the Jewish question later on in a free Poland. For his disobedience he was sent to prison.

A similar case is that of the distinguished clergyman Monsignor Marceli Godlewski who, as a Polish nationalist, was also a strong supporter of the National Democrats. An overt anti-Semite, Godlewski was eager to make his anti-Jewish position widely known. Many of his sermons and writings were filled with attacks on Jews.

With the Nazi occupation Godlewski's behavior underwent a drastic change. Joining the Jews in the Warsaw ghetto, he attended there to the spiritual needs of those who converted to Catholicism. When it came to more tangible aid, he made no distinction between Jews and converts. He organized a kitchen that fed the poor regardless of their religion. In his desire to help, Godlewski was particularly concerned about the fate of Jewish children. Some he fed, others he found shelter for in the forbidden Christian world in convents and Catholic orphanages.[4]

As much as it is fascinating to study the anti-Semitic helper, it is also important to put this unique situation into some perspective. The case of the anti-Semitic helper, while, no doubt, fascinating, is also extremely rare. In both the literature and archival materials, the same few names appear and reappear.[5] Also as I have noted earlier, of the thirty-one rescuers I interviewed directly, only two fit the definition of overt anti-Semites.

Indirect evidence bearing on this question of overall numbers is also available. The two largest political blocs in prewar Poland, the National Democrats and the ruling Sanacja party backed the existing anti-Jewish measures. Together, these two parties made up a solid majority in the Polish parliament.[6] Yet, of the 189 rescuers I studied, only 10 percent belonged to both of these two parties. Similarly, of the more than 500 helpers mentioned by survivors, only 4 percent were said to belong to one of these parties.

Polish testimonies and memoirs refrain from direct discussions and evaluations of Jews. Of the 189 rescuers, only 87 expressed any opinion about Jews. Only 15 percent of those who did, held negative attitudes. Similarly, the majority of those who commented about Jews expressed no anti-Jewish stereotypes.[7]

The rescuers' positive attitudes to Jews are further supported by a comparison of the paid and nonpaid helpers. As reported by the sample of survivors, the majority of paid helpers had negative attitudes toward Jews, while such explicit negative attitudes were mentioned by only a small minority of rescuers.[8] Although uneven and scattered, my evidence nevertheless shows a consistent pattern: among righteous Poles, anti-Semites were an exception.

Rare as these anti-Semitic rescuers seem to be, what characteristics, if any, do they have in common, and what prompted them to risk their lives for the persecuted Jews?

Of the Poles I interviewed directly, two were members of an anti-Semitic political party. Before the war, both did not hesitate to express anti-Semitic views and were clearly identified as anti-Semites. Marek Dunski, a well-known writer, made his anti-Jewish views known through his publications. A conservative and devout Catholic, he looked with disfavor on the Jewish presence in Poland, but this did not prevent him from saving Jews. As a high executive in the Department of Welfare, he took the responsibility for illegally placing Jewish children in orphanages and convents by signing orders for their acceptance. Dunski has been credited with the rescue of 500 Jewish children.

Even though Dunski risked his life for Jews, he retained his anti-Semitic attitudes and values. To illustrate, when I asked him what effects the disappearance of Jews had in Poland, Dunski remarked that the Jewish presence had in the past prevented the country from achieving unity, but now Poland was ethnically homogeneous. Although he himself was helping rather than harming Jews, he perceived their absence as a positive development.

Hela Horska was also socially prominent. She and her physician husband were both members of the anti-Semitic National Democratic party and both were devout Catholics. Still, they saved fourteen Jews and extended more temporary help to scores of others. Hela Horska's views of Jews remained surprisingly negative. Her opinion about Polish postwar anti-Semitism and the 1968 anti-Jewish purge was that it all happened for "political reasons." Pressed for a more precise explanation, she gave an anti-Semitic answer: *Israel needed soldiers and wanted the Polish Jews to emigrate. Therefore, the Jews themselves created anti-Jewish propaganda in Poland, which in turn scared the Jews and made them leave for Israel. In reality, once they got to Israel they hated it.*

Horska even blamed the Jews for her personal tragedy. As a young doctor, her son was eager to train abroad and chose Israel. The only opening offered to him was in cancer research. Horska was convinced that he developed a low white blood cell count as a result of his work. Had the Jews offered him work in another branch of medicine, he would have remained a healthy man. Ultimately, then, the Jews were to blame for his poor health.

Only after insisting that I stop taping did Horska bitterly complain about the ingratitude of the Jews she had saved. She nevertheless agreed to give me the address of David Rodman, the only one of them who has kept in touch with her.

In contrast to Horska's view of her charges, Rodman had the highest opinion of his rescuers, even though he knew about their anti-Semitism. While they did not conceal their anti-Jewish views, David emphasized that their anti-Semitism did not interfere with their strong compassion: *Once she [Hela Horska] and I were watching from the window how the Nazis were executing Jews. Before I realized what had happened she fell down and fainted. . . . She used to come home and cry for the Jews. She used to run around incessantly trying to place Jewish children, she even forgot to eat, she was so absorbed in helping.*

For a while their house was like a transit station. Anybody who came got food, money, and shelter for at least one night. Sometimes she would place a child, and the next day the people would bring it back. Someone might have recognized it, or they changed their mind. But she did not give up. Again she made all the effort to place it. . . . [The Horskis] felt that the presence of Jews in Poland was an economic and social problem, but they never dreamt of solving it by murdering Jews. They were highly regarded in the community. As far as I know all their friends were anti-Semites. I doubt it very much if any of his friends would have moved a finger for a Jew. I knew them well. I doubt if any of them helped at all. In fact, it was the hardest thing for a Pole to help a Jew. He had to rise almost like out from the stratosphere.

Some other rescuers and survivors I spoke to also tried to describe and explain the actions of anti-Semitic helpers. Thus, the survivor Vera Ellman told me about one such helper, Komornicka.[9] Another devout Catholic, and a prominent member of the conservative political party, the National Democrats, Komornicka had a high position in the Department of Education. Before the war she would come to inspect the school in which Vera worked. Unmarried, small, a disfigured hunchback, Komornicka was known for her virulent anti-Semitism. Vera remembered: *During the war I . . . went to live in Warsaw with Polish friends who protected me in a most gracious and considerate way. Three months later on the street I met Komornicka. Automatically full of fear, I tried to run away. She ran after me. When she caught up with me she grabbed me by the hand saying: "My child, you are running away from me?" I told her that I was well aware of her attitude to Jews, and of course I was trying to avoid her. Komornicka began to cry and begged me to come and visit her. When I refused, she gave me her card saying: "Just in case you need something I am leaving you my card." At home I told my Polish friends about this encounter. They scolded me for my rudeness, explaining that she was one of the finest people, a prominent underground figure who was helping Jews. They insisted that I call her and apologize. I followed their advice and told Komornicka where I was. She was very grateful, saying that God would reward us for trusting her. . . . Later on, after I became active in the Council for Aid to Jews, I used to meet her there. We became friends. At one of those meetings I remember her saying: "You know how religious I am, but if Christ would stand between me and Hitler, through Christ's body I would knife him down. . . ." When I wanted to become the director of a boarding school for 120 women, Komornicka, who was very influential, helped me get the job. She wrote a letter of recommendation in which she claimed to have known me for years and in which she guaranteed my suitability. Because of her I got the job. I was only one of the many Jews she helped. She was a lonely person. After the war when she became sick I took care of her. She died in my arms.*

Nearly all anti-Semitic rescuers that came to my attention shared a number of characteristics. They were devout Catholics, highly nationalistic, intellectually and socially prominent. Their rescuing of Jews, however, did not seem to change their anti-Semitic views. On the contrary, it was as if these anti-Semitic sentiments and values merely took a leave of absence. They remained surprisingly intact. How do these common characteristics and patterns relate to the rescuing of Jews?

Dunski and Horska explained their aid to Jews in religious terms. Dunski said: *My motivation had to do with my religious convictions. I did it as a Christian. One could not simply allow a person to die. . . . One had to do what had to be done. . . . I did what I did as a good Catholic.* In a slightly different way Horska essentially echoes his statement: *I am a deeply religious person and I believe that the world became so horrible, so cruel, that when we would have to account before God one would at least have to point to a few good deeds in this hell on earth. Someone had to help these persecuted Jews. And so they stayed with us.*

Yet both Dunski and Horska were good Catholics before the war. Before the war their religious convictions, far from preventing them from discriminating against Jews, might have contributed to their anti-Semitism. Their religion in itself does not explain the shift from condemnation to protection. Rather, in acting on behalf of Jews these anti-Semites were responding to the Nazi terror. It is likely, however, that their Catholic background made them more sensitive and aware of the Nazi crimes.

Some of those I interviewed tried to explain how devout Catholics who were also anti-Semites ended up protecting Jews. The rescuer Janka Polanska was relatively free of anti-Jewish prejudices. As a patriot she was concerned about Poland's image, and eager to show that her fellow citizens were opposed to the Nazi persecution of Jews. To support her views she talked about anti-Semitic rescuers, whose very presence, she felt, showed a general Polish rejection of Jewish extermination.

Polanska cited her friend, a prominent physician, an anti-Semite who saved Jews. His explanation was also related to religious motivations: *When I saw how the Germans were hitting Jews, how they were pushing them, when I saw all the dying and dead Jews around me, then I thought that Christianity was worthless, then I became convinced that I must save whomever I could.* This man considered the Germans inhuman and felt that by endorsing anti-Semitism he had indirectly contributed to the Nazi inhumanity: *And so in order to regain at least some of his humanity he had to save Jews. He had to sacrifice himself for the persecuted Jews in order to live with himself.*

Felicia Zapolska, a helper who admitted freely to her disappointment with Jews, did not quite fit the definition of an anti-Semite. Speaking about anti-Semitic helpers, she, too, saw their actions in religious terms: *They felt in a Catholic fashion, "here I considered a Jew as a low being. I was an anti-Semite, maybe I behaved improperly towards them and for this I must pay now. In some way I must erase this sin." It is in a deep Catholic sense that they tried to atone for their sins. I will pay for the sin, for having considered this human being as something inferior.*

Bolesław Twardy more specifically traced the rescuing of Jews by anti-Semites to their religious upbringing: *There is something in the Catholic upbringing which forces one to review one's position—to reconsider whether one is committing a sin. For an intelligent Catholic, the horrible, the most cruel, glaringly murderous behavior towards the Jews was a sin . . . and a shock. Up until then they failed to realize what anti-Jewish prejudice could lead to. Once they realized its consequences they began to reflect on their own position. Suddenly it occurred to them that through their anti-Semitism they might have participated in something which was at odds with their basic religious precepts. So to alleviate their guilt they began to save Jews.* This is how it was with many Catholics and many of my own friends. Their reasoning was as follows: *"I have participated in something evil. I must erase what I did. The only way in which I can hope to erase it is by helping Jews."*

Not everyone explained the rescuing by anti-Semites in religious terms, at least not in the conventional religious sense. Some moved from religious motivations into a more abstract moral plane, expressing the view that indirectly and symbolically any brand of anti-Semitism, including their own, might have played a part in the Nazi extermination of Jews. A simple question by an anti-Semitic woman who offered help to a Jew shows how some of them must have perceived the situation: *Who could have foreseen that anti-Semitism would lead to mass murder?* She was convinced that after the war anti-Semitism would be prohibited by law.[10]

Dunski explained his personal aid in religious terms but speculated that others might have been prompted by Poland's changing conditions: *The war made people more introspective. In time of peace one says I hate this one because he has a long nose, or he talks differently. During the war people tend to forget about such things. They start to look at more basic things. They tend to see a person as a human being. This is what happened with the Jews. They were not seen as Jews but as human beings. Only the war showed us that underneath all this difference and all this strangeness we are all the same. . . . It was a question of a common struggle. We all felt as a part of the same group.*

Eva Anielska, who stood up for Jews before and during the war, interpreted the actions of anti-Semitic rescuers in similar terms: *The 1939 war was a terrible tragedy for all Poles, a terrible shock. All Poles including the average person went through a terrible disappointment, and even I who as a Socialist was in opposition to the government, I too was bitterly disappointed by the actions of our leaders. Our world around us fell apart in five days. We all went through a lot. Maybe as a result of all these upheavals the Poles realized that their ideas about life, all kinds of things and values ought to be questioned. I am not saying that Poles stopped being anti-Semites during the war. All I am saying is that they might have realized that their ideology and beliefs do not work. The war was a shock and with it came a reexamination of one's views.*

You know when I was in Pawiak [prison] all of us there were destined to die. Every day at five o'clock in the morning the Gestapo, cursing and screaming, would enter our cell with a list of names and remove some of us. The others in my cell were worse off than I. The Gestapo was torturing their children in front of their eyes in order to get secrets out of them. We all began to share the feelings of being insignificant and that nothing at all mattered. In the face of this terrible tragedy we became better. It was as if we became one family. This might have happened to some Christian anti-Semites. Perhaps they began to feel that they were in a similar situation as the Jew and that it was only a question of time. This feeling together with the doubts and the questioning of old values might have pushed them toward extending help.

David Rodman's assessment of the reasons his anti-Semitic protectors saved Jews resembles much of what others have said: *Their actions were motivated by very high moral values. You see, I cannot say that they helped us for personal reasons because they liked us. They tried to help anybody. . . . At one point the doctor was traveling by train. A German officer came to his compartment saying, "Get up you Polish swine." The doctor, a prominent citizen, had to get up and leave. He must have realized, perhaps remembered, that Jews were treated like this in prewar Poland. He might have realized that what is now happening to Jews can and does happen to them.*

It was not our presence or our relationship to them that changed their anti-Semitic views. Instead, it was the German actions that made them change, by upsetting all preexisting values including those which had to do with anti-Semitism.

When faced with Nazi atrocities, some of these intellectual anti-Semites began to blame themselves for Nazi crimes against Jews. To repent for what they thought they might have indirectly contributed to—namely, their own anti-Semitism—they tried to save Jews.

Helping a person and having a positive attitude about this person do not necessarily go hand in hand. One survivor explained: *It is one thing to help an underdog and quite another to consider him as an equal. As long as the Jews were persecuted one could help them, but when they wanted to become equal to the Christian, this is quite a different matter. It is one thing to help someone who is down and quite another to consider the same person as one's equal.*

When trying to understand the rescuing of Jews in general, and particularly by anti-Semites, some have pointed to another wartime development. In prewar Poland Jews were perceived as dangerous and aggressive. Nationalist Poles, more so than others, defined Jews as a threat to their country's unity. Polish nationalism then went hand in hand with anti-Semitism, and it can be safely assumed that most nationalists were also anti-Semites.[11]

The Nazi treatment of the Jews, however, transformed them into helpless beings condemned to death. The intellectual Poles in particular had to become aware of this change.

Dunski put it this way: *During the war anti-Semites did not see a Jew as a Jew. He did not anymore fit the image of the threatening aggressive foreign being. Instead, the Jew assumed the position of the haunted and the persecuted. He was stripped of all attributes except those of a hurt, suffering being. The anti-Semite saw him as such and responded to the appeal of humanity.*

In a sense then the transformation of the Jew into a totally dependent being, desperately in need of protection, made it easy for some anti-Semites to act in opposition to their prejudices. This transformation overshadowed what they had felt were the "Jewish" characteristics.

In addition, it seems that many of the anti-Semitic rescuers had a history of helping the needy. In fact, already before the war, most of the other anti-Semitic rescuers had engaged in charitable work. Their prewar help to the poor did not include Jews because at that time the Jews were perceived as threats and not as needy. Yet the Nazis through their policies of extermination transformed the image of the Jew from a competitive, aggressive being into a suffering, deprived, and totally helpless individual. This change in turn was compatible with their predisposition to protect the underdog.

The last example, that of Zofia Kossak-Szczucka, seems particularly exceptional and dramatic. Thus, Zofia Kossak-Szczucka came from a socially prominent Polish family. An established writer of historical novels, a devout Catholic, a member of the Catholic organization (FOP), she was also a nationalist known for her rightist views. None of these characteristics had qualified her for becoming the defender of Jews. And yet, as an active member of the underground, in the name of this Catholic organization, in the summer of 1942 Zofia Kossak-Szczucka wrote an illegal leaflet: "The Protest."[12] The stated aim of this document was to protest the Nazi crimes committed against the Jews. In reality this leaflet is a strange mixture of outrage against German atrocities, indignation against the indifference of the Free World, and a reiteration of anti-Jewish sentiments.

As a representative of the Catholic organization (FOP), Zofia Kossak-Szczucka followed up "The Protest" with a request to establish an underground organization to save Jews. By the end of 1942 the Council for Aid to Jews became a reality.*

Although she never became an official member of this organization, Szczucka continued her aid to Jews and also her other underground work. In 1943 she was caught and sent to Auschwitz. After her release nearly a year later, she resumed her work becoming especially active in the rescue of Jewish children by hiding them in convents and other institutions run by the clergy.[13]

Rare, surprising, and intriguing, Zofia Kossak-Szczucka's case also begs for an explanation. Surely, she must have been moved by the Jewish plight, but this does not explain much. How did she reconcile her unsympathetic view of Jews with her selfless efforts on their behalf?

Rather than speculate, I will let Zofia Kossak-Szczucka answer through one of the characters she created. This is a conversation between two Polish men who are friends; both agree on the pervasive anti-Semitic views about Jews. Both are avid anti-Semites and conscious of it, but only one of

---

*For the actual document and translation, see end of this chapter.

them is protecting Jews. The Pole who does not aid Jews asks for an explanation. This is what he hears:

Today the Jews face extermination. They are the victims of unjust murderous persecutions. I must save them. "Do unto others what you want others to do unto you." This commandment demands that I use all the means I have to save others, the very same ways that I would use for my own salvation. To be sure, after the war the situation will be different. The same laws will apply to the Jew and to me. At that point I will tell the Jew: "I had saved you, sheltered you when you were persecuted. To keep you alive I risked my own life and the lives of those who were dear to me. Now nothing threatens you. You have your own friends and in some ways you are better off than I. Now I am depriving you of my home. Go and settle somewhere else. I wish you luck and will be glad to help you. I am not going to hurt you, but in my own home I want to live alone. I have a right to it. . . ." Christ stands behind every human being and watches how we react to Him. He stretches His hand to us through a runaway Jew from the ghetto the same way as He does through our brothers. He is close to everyone."[14]

Like other such helpers, through her aid she tried to disassociate herself from the crimes committed against the Jews. This disassociation, however, in no way changed her views about them.

Precisely because it was least expected, the rescuing of Jews by anti-Semites has attracted a great deal of attention. Undue focus on such helpers gave the impression that this behavior was common. In reality, however, although frequently discussed such anti-Semitic rescuers are as rare as they are unusual. The fact that among rescuers anti-Semites were a rare exception underlies the basic incompatibility between anti-Semitism and the risking of lives for Jews.

Certain resemblances among these few anti-Semitic rescuers point to some tentative conclusions. Most of them were devout Catholics: all were highly nationalistic, intellectual, and socially prominent. The Polish Catholic church had for centuries been both strongly anti-Semitic and highly nationalistic. Catholicism, nationalism, and anti-Semitism were at times associated with social prominence and intellectualism. Of the three, nationalism was particularly compatible with social prominence and intellectualism. Being an intellectual implies an ability to rethink critically and reevaluate situations. It is therefore no surprise that some intellectuals, when faced with the startling changes the war brought about, began to question their personal values. Some were willing to concede that the Germans were a greater threat than the Jews; thus their devotion to Poland, especially during the Nazi occupation, overshadowed most other considerations.

Moreover, appalled by the Nazi atrocities, a few nationalistic anti-Semites were eager to disassociate their country and themselves from Nazi crimes. One way in which they could do that was by saving Jews. Paradoxically, whereas before the war Polish nationalism was in part responsible for

Polish anti-Semitism, during the war—together with other special circumstances—it could lead to the protection of Jews.

More significantly, this special group of anti-Semites began to believe that anti-Semitism in general—including their own—had to bear some indirect responsibility for the wartime cruelty and devastation. With this realization came a burden of personal guilt for the Jews plight, and with it a desire to repent.

Those few who experienced the desire to repent found it easier to act upon their feelings because the Nazi policies of destruction transformed the Jews from seemingly aggressive, threatening people into helpless and suffering beings in need of protection. Finally too, in and of itself, this transformation had appealed to the already existing compassion for the poor.

To be sure, even as they acted on behalf of Jews, these rescuers knew that their brand of anti-Semitism was less pernicious, but they were willing to put it aside only temporarily.

To reflect on the meanings and implications of one's values and actions requires a certain measure of independence and sophistication. It is probably no coincidence that all anti-Semitic rescuers were socially prominent. Nor is it an accident that most of them were devout Catholics. The Catholic religion demands of its followers a reassessment of values. The option to do this, however, is usually open to the intellectual Catholics. That is, rather than follow the religious precepts in a narrow, concrete way, these anti-Semitic rescuers reflected on their meaning and implications.

In view of the many complicated forces that had to be overcome to save Jews, it is not surprising that among Polish rescuers overt anti-Semites were but a rare exception. Precisely because rescue by anti-Semites was rare it remains an intriguing exception.

P R O T E S T !

W ghetcie warszawskim, za murem odcinającym od świata, kilkaset tysięcy skazańców czeka na śmierć. Nie istnieje dla nich nadzieja ratunku, nie nadchodzi znikąd pomoc. Ulicami przebiegają oprawcy, strzelając do każdego, kto się ośmieli wyjść z domu. Strzelają podobnie do każdego, kto stanie w oknie. Na jezdni walają się niepogrzebane trupy.

Dzienna przepisowa ilość ofiar wynosi 8—10 tysięcy. Policjanci żydowscy obowiązani są dostarczyć ich do rąk katów niemieckich. Jeżeli tego nie uczynią, zginą sami. Dzieci nie mogące iść o własnych siłach są ładowane na wozy. Ładowanie odbywa się w sposób tak brutalny, że mało które żywe dojeżdża do rampy. Matki patrzące na to dostają obłędu. Ilość obłąkanych z rozpaczy i grozy równa się ilości zastrzelonych.

Na rampie czekają wagony kolejowe. Kaci upychają w nich skazańców po 150 osób w jednym. Na podłodze leży gruba warstwa wapna i chloru polana wodą. Drzwi wagonu zostają zaplombowane. Czasem pociąg rusza zaraz po załadowaniu, czasem stoi na bocznym torze dobę, dwie... To nie ma już dla nikogo żadnego znaczenia. Z ludzi stłoczonych tak ciasno, że umarli nie mogą upaść i stoją nadal ramię w ramię z żyjącymi, z ludzi konających zwolna w oparach wapna i chloru, pozbawionych powietrza, kropli wody, pożywienia — i tak nikt nie pozostanie przy życiu. Gdziekolwiek, kiedykolwiek dojadą śmiertelne pociągi — zawierać będą tylko trupy...

Wobec tej męki wyzwoleniem stałby się rychły zgon. Oprawcy to przewidzieli. Wszystkie apteki na terenie ghetta zostały zamknięte, by nie dostarczyły trucizny. Broni nie ma. Jedyne co pozostaje, to rzucenie się z okna na bruk. To też bardzo wielu skazańców wymyka się katom w ten sposób.

To samo co w ghetcie warszawskim, odbywa się od pół roku w stu mniejszych i większych miasteczkach i miastach polskich. Ogón na liczba zabitych żydów przenosi już milion, a cyfra ta powiększa się z każdym dniem. Giną wszyscy. Bogacze i ubodzy, starce, kobiety, mężczyźni, młodzież, niemowlęta, katolicy umierający z Imieniem Jezusa i Maryi, równie jak starozakonni. Wszyscy zawinili tym, że się urodzili w narodzie żydowskim, skazanym na zagładę przez Hitlera.

Świat patrzy na tę zbrodnię, straszliwszą niż wszystko, co widziały dzieje i — milczy. Rzeź milionów bezbronnych ludzi dokonywa się wśród powszechnego, złowrogiego milczenia. Milczą kaci, nie chełpią się tym co czynią. Nie zabierają głosu Anglia ani Ameryka, milczy nawet wpływowe międzynarodowe żydostwo, tak dawniej przeczulone na każdą krzywdę swoich. Milczą i Polacy. Polscy polityczni przyjaciele żydów ograniczają się do notatek dziennikarskich, polscy przeciwnicy żydów objawiają brak zainteresowania dla sprawy im obcej. Ginący żydzi otoczeni są przez samych umywających ręce Piłatów.

Tego milczenia dłużej tolerować nie można. Jakiekolwiek są jego pobudki — jest ono nikczemne. Wobec zbrodni nie wolno pozostawać biernym. Kto milczy w obliczu mordu — staje się wspólnikiem mordercy. Kto nie potępia — ten przyzwala.

Zabieramy przeto głos my, katolicy - Polacy. Uczucia nasze względem żydów nie uległy zmianie. Nie przestajemy uważać ich za politycznych, gospodarczych i ideowych wrogów Polski. Co więcej, zdajemy sobie sprawę z tego, iż nienawidzą nas oni więcej niż Niemców, że czynią nas odpowiedzialnymi za swoje nieszczęście. Dlaczego, na jakiej podstawie — to pozostaje tajemnicą duszy żydowskiej, niemniej jest faktem nieustannie potwierdzanym. Świadomość tych uczuć jednak nie zwalnia nas z obowiązku potępienia zbrodni.

Nie chcemy być Piłatami. Nie mamy możności czynnie przeciwdziałać morderstwom niemieckim, nie możemy nic poradzić, nikogo uratować, — lecz protestujemy z głębi serc przejętych litością, oburzeniem i grozą. Protestu tego domaga się od nas Bóg, Bóg który nie pozwolił zabijać. Domaga się sumienie chrześcijańskie. Każda istota, zwąca się człowiekiem, ma prawo do miłości bliźniego. Krew bezbronnych woła o pomstę do nieba. Kto z nami tego protestu nie popiera — nie jest katolikiem.

Protestujemy równocześnie jako Polacy. Nie wierzymy, by Polska odnieść mogła korzyść z okrucieństw niemieckich. Przeciwnie. W upartym milczeniu międzynarodowego żydostwa, w zabiegach propagandy niemieckiej usiłującej już teraz zrzucić odium za rzeź żydów na Litwinów i... Polaków, wyczuwamy planowanie wrogiej dla nas akcji. Wiemy również, jak trujący bywa posiew zbrodni. Przymusowe uczestnictwo narodu polskiego w krwawym widowisku spełniającym się na ziemiach polskich, może snadno wyhodować zobojętnienie na krzywdę, sadyzm i ponad wszystko groźne przekonanie, że wolno mordować bliźnich bezkarnie.

Kto tego nie rozumie, kto dumną, wolną przyszłość Polski śmiałby łączyć z nikczemną radością z nieszczęścia bliźniego, — nie jest przeto ani katolikiem, ani Polakiem!

FRONT ODRODZENIA POLSKI

Translation

The Protest!

In the Warsaw ghetto, behind a wall that is cutting them off from the world, several hundred thousand condemned people await death. No hope of survival exists for them, and no help is coming from anywhere. The executioners run through the streets shooting anyone who dares leave the house. They shoot anyone who is near the window. The streets are full of unburied corpses.

The prescribed number of victims is 8,000–10,000 daily. Jewish police must deliver the victims to the hands of the German executioners. If they do not, they themselves perish. Children cannot walk on their own strength and are loaded into wagons. The process of loading is so brutal that very few survive it. Mothers, looking on, become insane. The number of people insane from grief and horror equals the number of people shot to death.

Railroad cars wait on the ramp. The executioners pack the condemned into the wagons, 150 of them in one wagon. On the floor of the wagon lies a thick layer of lime and chloride poured over with water. The door of the wagon gets sealed. Sometimes the wagon moves immediately after loading, sometimes it stays on the siderails a day, two—it doesn't matter to anyone anymore. Of the people crammed so tightly that the dead cannot fall and keep standing arm in arm with the living, of the people slowly dying in the fumes of the lime and chloride, being deprived of air, water, food—no one will remain alive anyway. Whenever, wherever the death-wagons will arrive, they will only contain corpses.

In the face of this suffering, speedy death would be liberation. The executioners foresaw this: all pharmacies in the ghetto were closed to prevent the delivery of poison. There are no weapons. What is left is jumping from a window onto the pavement, and thus many condemned escape their executions in this manner. Just as in the Warsaw ghetto, since six months ago in larger and smaller Polish towns and cities the same is happening. The total number of Jews killed is over one million, and this number is growing daily. All perish: the rich and poor, the old, the women, the men, the youth, the babies. The Catholics dying, with the name of Jesus and Mary, like the Jews: all were guilty of being born as Jews, who were sentenced to annihilation by Hitler.

The world looks upon this murder, more horrible than anything that history has ever seen, and stays silent. The slaughter of millions of defenseless people is being carried out amid general sinister silence. Silent are the executioners; they do not boast about their deed. England and America are not saying anything. Silent is the ever-influential international Jewry, which was previously oversensitive of wrongdoing to their own. Silent are Poles. Polish political friends of Jews limit themselves to newspaper notes; Polish opponents of Jews show lack of interest in the problem, which is foreign to them. The perishing Jews are surrounded by Pilates who deny all guilt.

This silence can no longer be tolerated. Whatever the reason for it, it is vile. In the face of murder it is wrong to remain passive. Whoever is silent witnessing murder becomes a partner to the murder. Whoever does not condemn, consents.

Therefore we—Catholics, Poles—raise our voices. Our feeling toward the Jews has not changed. We continue to deem them political, economic, and ideological enemies of Poland. Moreover, we realize that they hate us more than they hate the Germans, and that they make us responsible for their misfortune. Why, and on what basis, remains a mystery of the Jewish soul. Nevertheless, this is a decided fact. Awareness of this fact, however, does not release us from the duty of damnation of murder.

We do not want to be Pilates. We have no means actively to counteract the German murders; we cannot help, nor can we rescue anybody. But we protest from the bottom of our hearts filled with pity, indignation, and horror. This protest is demanded of us by God, who does not allow us to kill. It is demanded by our Christian conscience. Every being calling itself human has the right to love his fellow man. The blood of the defenseless victims is calling for revenge. Who does not support the protest with us, is not a Catholic.

We protest also as Poles. We do not believe that Poland could benefit from the horrible Nazi deeds. Just the opposite; we detect hostility toward us, caused by the silence of world Jewry and by the German propaganda, already in process to shift from themselves the blame for the slaughter of Jews to Lithuanians and—Poles.

The forced participation of the Polish nation in the bloody spectacle taking place on Polish soil may breed indifference to unfairness, sadism, and, above all, belief that murder is not punishable.

Whoever does not understand this, and whoever dares to connect the future of the proud, free Poland, with the vile enjoyment of your fellow man's calamity—is, therefore, not a Catholic and not a Pole.

THE FRONT FOR THE REBIRTH
OF POLAND

PART III

Explanations of Righteous Rescue and Rescuers

CHAPTER 7

The Influence of
Class and Politics

Here, in this third and final part of the book, I concentrate on the "whys" rather than the "whats," on explanations rather than descriptions. Whatever findings emerge may reach beyond the Holocaust. They may tell who in time of trouble would be more likely to stand up for the defenseless and the persecuted.

To unravel the characteristics and motivations of the righteous, I start by considering currently held assumptions, continue by adding new observations, and end by summarizing and integrating the material of this book. In the next three chapters, then, I will examine the extent to which class, politics, friendship, and religion identify the rescuers and explain their protection of Jews.

Traditionally class and politics have functioned as important predictors for behavior. When it comes to class and rescue, some have assumed that righteous Christians are not distinguishable in terms of social class,* that they come from all levels of society.[1] Others, trying to describe the occupational affiliation of helpers, have similarly pointed to a wide array of categories, ranging from celebrated writers to grave diggers.[2] Still others who have tried to develop special typologies have concluded that it is difficult to differentiate helpers from the rest of the population.[3]

Among those willing to make more specific predictions, a central assumption has been that the working or lower class showed a greater propensity to help than others.[4] Thus, noting that in every nation there are people who put themselves out for others, Zygmunt Rostal, a rescuer who came from a poor laborer's background, conceded only very reluctantly that other groups too might have helped, but emphasized that *among the laborers, in my circle, one rarely saw people who were completely indifferent*. In Zygmunt's view most working class people were sympathetic to the Jewish plight and hence were more likely to protect them.

*"Class" is here defined loosely by occupation

Did being a worker significantly influence one's need or desire to help? If it did, what was the reason for this increased sensitivity? As a Socialist, the rescuer Eva Anielska felt that lower class people, themselves poor and socially deprived, were in a better position to identify with the suffering Jews, and hence more likely to reach out to anyone in need.

Although the survivor Pola Stein at first insisted that helpers could not be distinguished from others, later on she too explained: *I believe that the poor people were better than the rich. . . . Why? They have more natural instincts. When I spoke to Jews who survived, those who helped them were on the whole uneducated, simple people.*

Are these occasional statements supported by more systematic evidence? What, in fact, was the proportion of workers among rescuers? The numerical evidence yields interesting results. Describing their helpers in terms of class, Jewish survivors placed a minority (21 percent) into the working class category.[5] The same figures emerge when the Polish rescuers themselves are looked at in terms of class.[6] When paid helpers and rescuers are compared in terms of class, the proportion of those who helped exclusively for money is slightly higher (27 percent). Small as this difference is, its direction, in particular, casts doubt on the idea that workers had a greater propensity for selfless protection of Jews than others.

One additional factor should be noted. In prewar Poland approximately 20 percent were considered working class, precisely the percentage of workers among rescuers. Thus, my evidence seems to suggest that laborers were not more likely than Poles from other social classes to save Jews.

Indeed, not all who touched on the issue of class gave special credit to workers. Some saw the Polish intelligentsia, particularly those active in the underground, as being most active in the rescuing of Jews. Proponents of this view cite many celebrated writers, journalists, and doctors who risked their lives for Jews.[7]

While still believing that laborers were most likely to help, Eva Anielska was willing to admit that many of the other rescuers were intellectuals.* She further specified that *among the intellectuals, those who helped the most were people who worked for social causes.*

In prewar Poland, only 10 percent of the population were intellectuals.[8] The survivors placed 19 percent of their rescuers into the category of intellectuals.[9] Even more startling, this figure nearly triples when the sample of Polish rescuers is looked at directly, increasing to slightly over 54 percent.[10]

How should this sharp jump be interpreted? It is important to realize that direct information about rescuers comes from published memoirs and testimonies that the rescuers offered to historical commissions. Individuals who were members of the intelligentsia may have simply made up a larger percentage of those willing to write about or recount their experiences,

---

*The terms "intellectuals" and "intelligentsia" are here used interchangeably. Each applies roughly to professionals and those in the creative fields.

since both writing and even giving oral testimony are skills related to education. Thus I cannot accurately ascertain to what extent this large proportion of intellectuals in the rescuer sample may reflect a real situation and how much of it is due to a sample bias.

The sample of Jewish survivors may, like that of the Polish rescuers, contain a disproportionate number of the better educated. Still, when describing their helpers they would have no reason to omit any special group from their descriptions. The survivors' statements about their Polish rescuers therefore might be more representative than the direct evidence from rescuers.

As a precaution, then, when considering proportions of intellectual rescuers, I will rely on the less dramatic figures derived from Jewish survivors' reports. Nonetheless, even these lower figures are impressive. Lending further support to the notion that intellectuals made a more than representative contribution to selfless rescue is the fact that only three percent of the paid helpers were intellectuals.[11]

Next, we come to the peasants. Since in prewar Poland, the peasants were the largest single class and in sheer numbers make up the largest group of rescuers and of paid helpers, they have been the subjects of a great deal of attention. By some they were described as noble, by others as cruel, with the bulk of opinions leaning toward the negative. Earlier I have shown that some peasants were executed for their selfless help, while others were busy rounding up Jews and delivering them to the authorities.[12] Some Poles were appalled by the peasants' behavior, the special raids, the hunting of Jews like wild animals. Frequently, too, the peasants did not even bother to deliver their victims to the authorities, but would kill them instead.[13]

Lola Freud was one of the many survivors who were affected by the atrocities of the Polish peasants. She was working for a farmer who did not know that she was Jewish. Eventually she told him who she was when she asked him to hide her two brothers. For money the peasant granted her this request. While she moved around freely, passing as a Pole, her two brothers stayed in a specially built hole in the fields. Each day at dusk one of the brothers would sneak into the barn for food. Once, instead of returning, he stayed overnight in the barn. Before dawn, brother and sister could hear shots from the direction of the field. Through the barn's opening they saw peasants walking away from the fields. Lola recalled: *I said to myself, I hope nothing happened to my brother. For safety we had to wait for an hour. Then my brother and I went into the fields. I stood guard to make sure that no one saw us, while my brother moved closer to the hiding place. It was a small place built just for two. Then I heard my brother's desperate scream: "Solomon is dead, Solomon is dead." In a split of a second I reached his side. All I could see was blood, I could not see my dead brother, only blood. The entire hole was full of blood. We grabbed each other's hands and started running. We were just running, we didn't know where we were running. We ran through the fields, forests, we did not know where we were going, and what we were doing. We ran and ran and ran . . .*

Like Lola, Frida Nordau survived because of peasants' help, but she was also subjected to their cruelty and denouncements. Frida succeeded in overcoming the many close calls and perils, while her mother was less fortunate.

At one point Frida's family had to split into two groups. They were to meet when it was safer. But instead of meeting, this is what happened: *When they (the other group including her mother) went in the opposite direction from ours, they were caught by a band of peasants. They tied them up like cattle, put them on wagons, and brought them to a place five kilometers from here. There they delivered them to the Germans. They were all shot: my mother, my brother, his teenage son, and my nine-year-old sister.*

The cruelty of Polish peasants was described and deplored by many rescuers as well. One of these was Stach Kaminski, the Polish diplomat who, during the war, worked tirelessly on behalf of Jews. At first he saw no specific differentiation between those who saved Jews and those who did not.

Later, however, Kaminski reversed his position: *I must say with sorrow that the worst situation as far as helping Jews was concerned was in the countryside. There a Jew was a strange being.*

He continued to tell how after the war he visited the countryside and spoke to many peasants who admitted that they had caught Jews in the forest and delivered them to the local authorities. Asked why they did they explained that it was an act of mercy. The Jews, these peasants felt, lived in horrible conditions and bore little resemblance to human beings. By delivering Jews to the Germans they were relieving them of their misery. Kaminski concluded: *The Polish peasants were very backward. They had little contact with Jews and therefore Jews were strangers to them. Only some among them were hiding Jews for material gains which helped them survive economically.*

Are these illustrations supported by more systematic evidence? Of the Jewish survivors, 42 percent reported being saved by peasants. Among the rescuers themselves, the proportion of peasants dwindled considerably,[14] although here, as with intellectuals, a sample bias may be distorting our figures. In sharp contrast to these two figures, the proportion of peasants among the paid helpers rose to a solid majority: 66 percent.

Given the fact that in prewar Poland peasants made up 60 percent of the population, according to most estimates, it seems that peasants were less inclined toward Jewish rescue than either the intellectuals or the working classes.

Those who have commented about the middle class and its role in the rescuing of Jews show a surprising amount of agreement. Almost all are convinced that the middle class failed to stand up for Jews.* Some explain

*In this context the amorphous term "middle class" refers to businesspeople, shopkeepers, and those who owned any kind of business.

that the Nazi solution to the Jewish problem allowed the Polish middle class, who were indebted to Jewish bankers and tradespeople, to rid themselves of their creditors.[15] Ringelblum explained that *for the anti-Semitic middle class to hide Jews . . . was out of the question; it was in their interest to get rid of the Jews.*[16]

Again, do my findings support this conclusion? Only a small minority of the Jewish survivors (3 percent of the sample) claimed that those who protected them were store owners and businesspeople. Among the rescuers the proportion of businesspeople is very similar at 4 percent. Among the thirty-one Polish helpers I interviewed, two were well-to-do businessmen. On the other hand, of the paid helpers, only one was identified as middle class. This comes as no surprise, since the wealthy would have no reason to risk their lives for money.

Nonetheless the proportion of middle class rescuers is practically the same as the proportion of middle class in the population at large. It appears that among Polish rescuers the middle class was not underrepresented.[17]

How then is class related to Jewish rescue? While the intellectuals were more apt to rescue, the peasants were least likely to participate in such selfless behavior. The working and middle classes show no special tendencies. Even though proportionately fewer peasants were saving Jews, being a peasant did not necessarily mean that an individual would not become a rescuer. Indeed, a substantial number of Polish helpers were peasants. Similarly, while intellectuals were overrepresented among the rescuers, being an intellectual by no means propelled a person toward this kind of help. Class, then, at best appears as only a weak predictor of Jewish rescue.

Turning to the next issue, how does political involvement relate to rescue? Were the politically minded more likely to risk their lives for Jews?

In wartime Poland, political involvement usually meant opposition to the Nazis. Some have argued that the politically alert were more likely to rescue Jews. Through such courageous acts, Poles could express their opposition to Nazi policies of subjugation and terror.

In my discussion of anti-Semitic rescuers I have shown that, for some, resentment toward the occupying forces overshadowed all other considerations. Thus they set aside their anti-Jewish prejudices and focused instead on their opposition to the Germans. As they did they also helped Jews. Of the two anti-Semitic rescuers I interviewed, Marek Dunski showed his awareness about political aspects of rescue by making statements like: *All of us were united by the common struggle against the enemy. . . . We all felt as a part of the same column—the same group.*

Other rescuers also viewed their protection of Jews as an act of political defiance. Dr. Adam Estowski was absorbed in illegal underground activities part of which had to do with the saving of Jews. He explained that *there was a certain united front against the Nazis. And if they destroyed the Jews this meant that one had to protect them.*

Wojtek Kominek's reasons for saving Jews were also partly political. As an underground fighter, he aided Jews because the then head of the Polish government in exile, Sikorski, ordered him to do so. Finally, even many rescuers who did not belong to any underground group perceived their help to Jews as an opposition to the enemy. One helper explained: *I feel good that I cheated the damned Germans. I fooled them. They were smart but not smarter than I and could not destroy all.*

How important were such political and patriotic motivations in the rescue of Jews? When the entire group of Polish rescuers is considered, more than half of them (55 percent) saw their protection of Jews as a protest against the enemy. Such perceptions, however, were mixed with other reasons and motivations. Practically all rescuers (95 percent) felt that their aid was motivated by the victims' suffering. Inevitably, then, the significance of rescue as a political protest pales when compared to the overwhelming majority of those whose protection of Jews was a reaction to their personal plight and hence an act of compassion.

Survivors' reports about the politics of those who protected them are very similar. While 41 percent felt that rescuers were motivated by patriotism, an overwhelming majority (81 percent) said that compassion and pity was responsible for the helpers' aid.[18]

Was membership in the Polish underground related to rescue? Half of the helpers in my study participated in illegal underground activities, and half did not. In addition, fewer than 25 percent of Jewish survivors made any reference to the politics of their helpers. With this limited evidence I tentatively conclude that, even though political involvement per se might have had a slight effect on rescue, this effect could not qualify as an explanation.[19]

Rather than dealing with political involvement in general, most discussions on the topic are concerned with specific political indentifications and their effect on peoples' willingness to save Jews.

It is a historical reality that in prewar Poland most leftist groups failed to support the anti-Semitic measures condoned by the government and initiated by the extreme political right.[20] This led to the widely held assumption that Communists, Socialists, and others with leftist leanings would be most likely to help Jews. Adding to the confusion, "at the Conference of the Polish Socialist Party held in Cracow in 1941, the majority rejected active opposition to Nazi depredation against the Jews. As a result, representatives of the two Jewish Socialist parties . . . left the conference in protest. However, local Socialist organizations, trade unions, and individual Socialists established contact with Jewish groups in the ghettoes."[21] What then can we say about the left?

First, and perhaps most important, is the fact that far from being homogeneous, the political left, both before and during the war, was composed of many subgroups that were competitive and at times even hostile to each other. For example, the Socialist party contained many factions, whose ideologies varied, as did their level of hostility to Communists and others. Similarly, the peasants, who represented a powerful political block, differed greatly in terms of their leftist and nonleftist ideologies.[22]

Finally, too, questions about political leanings and Jewish rescue are complicated by factors related specifically to the Polish underground. The complex, contradictory relationship between the Polish underground and the Jews remains a highly controversial topic.

As I pointed out earlier, the Council for Aid to Jews (Żegota) was created largely through the efforts of a member of the Catholic organization FOP— Zofia Kossak-Szczucka, a well-known writer also known for her rightist political views.

Her efforts to aid Jews were supported by the Socialist and democratic parties and eventually by the Polish underground representative, Delegatura. The Polish government in exile also gave its official approval.[23]

In the end, political views of those on the council ranged from right to left. Included were also Jewish representatives—from the Socialist party, the Bund (the Jewish workers party), and different Zionist groups organized into a Jewish National Committee. Excluded were the extreme right and the extreme left, the Communists. The extreme right refused to participate, while the extreme left was excluded at the insistence of the official representatives of the Polish underground. Their wishes had to be obeyed because all financial support from abroad had to pass via their organization.[24] Funds for the Council for Aid to Jews came in part from the Polish government in London and in part from Jewish organizations in the Free World. Both the actual and relative contributions of each remain a controversial topic.[25]

Throughout Żegota's existence (1942–1945) its members never ceased to request more money, insisting that the sums reaching them were inadequate. Yet, the fact is that as the organization grew, its budget also expanded.[26]

The most active role in Żegota was assumed by the Socialists (PPS), with Julian Grobelny as head. A representative of the Bund named Leon Feiner filled the second place. Representing the Peasant party (SL), Tadeusz Rek became another vice-president. Adolf Berman, as a representative of the Jewish National Committee, filled the post of secretary, and Ferdynand Arczyński, representing the Democratic Party (SD), became the treasurer.[27] The Catholic movement (FOP) sent as its representative Władysław Bartoszewski, known for his varied and continual aid to Jews.[28] With headquarters in Warsaw, Żegota soon established additional centers, in Lwów, in Cracow, and in Lublin.

The council's broad aim to save Jews was subdivided into more specific forms of help. As a rule, Jews escaping the ghettoes arrived in the Christian world without adequate financial resources. For a large number of them who could not risk working, the council offered a modest monthly pension. Although the sums distributed were meager, barely covering minimal expenses, they nevertheless prevented starvation.

The organization also smuggled money into ghettoes with the explicit purpose of keeping the inmates alive or helping them escape. In addition, the council undertook the manufacture and distribution of forged documents for those who had moved to the Christian world or were about to do so.

One other central activity involved finding shelter and escorting Jews either from the ghetto or between different places on the Christian side. A small section of the organization also offered medical help, particularly to Jews in hiding.[29]

Finally, a separate unit dealt exclusively with issues related to Jewish children. According to one estimate, this unit succeeded in saving 2,500 children. According to another estimate, altogether 40,000 to 50,000 Jews benefited from Żegota's help. This last figure apparently refers not to the actual number saved but rather to those who benefited in some way from the council's activities.[30]

Still, the relationship between the Polish underground and the Jews was by no means confined to the activities of the Council for Aid to Jews; nor did it entail only positive features. The very complexity of the Polish underground made many kinds of intricate relationships possible. A brief sketch of the underground may throw some light on these complex issues.

In 1939, after Germany's takeover, Poles formed a government in France that subsequently moved to London. Included in this new government were representatives of the four major political blocks: the Socialists (PPS), the Peasant party (SL), the Centrist party (SP), and different kinds of nationalists. Each of these blocks was in turn subdivided into a variety of subgroups, each reflecting a wide range of ideological shadings. Precisely because the Polish government in London was made up of so many competing, and even hostile political entities, it had difficulty formulating clear-cut policies. At best this government achieved a broad, loosely connected, and fragile kind of unity.

In occupied Poland the official armed force of the government in exile was the illegal underground known as the Home Army (AK). The Home Army, like its political counterpart in London, was a conglomeration of many groups, also reflecting a wide range of political ideologies, each taking its cues from a different political segment in London. The entire AK was made up of semiautonomous units that often disagreed with each others' policies.

Excluded both from the Polish government in exile and from the Home Army in Poland were two small political entities, one the extreme right and the other the extreme left. The extreme right established its own underground, which was also composed of loosely connected splinter groups. The best known among them were the National Armed Forces (NSZ) and the Fascists, Falanga. Some of these groups fought the Germans, the Jews, the Communists, and factions of the Home Army (AK).[31]

In contrast to these fiercely independent rightist groups, the Communist Polish underground was largely dependent on the Soviets. This dependence denied them a broader acceptance from the Polish population.

Russian–Polish hostilities had a long tradition, reaching into the historical beginnings of each country. In the more recent past, for over 100 years, a large part of Poland had been ruled by a succession of oppressive Russian czars. Only with the end of World War I, when Poland reemerged as an independent nation, did it reclaim these occupied territories. But hostilities toward the Russians did not disappear with the end of the occupation. In Poland all Russian influence was highly suspect. No doubt because of its connection to Russia, few Poles found Communism appealing. In 1938, when the Polish government outlawed the Communist party, it had a membership of 20,000. Following the introduction of this law some Polish Communists were imprisoned; some escaped to the Soviet Union, and still others went into hiding.[32]

Because of the Russian–German alliance the handful of Polish Communists at first had to remain inactive. Only after the German attack on Russia in June 1941 could they start organizing into an anti-Nazi unit. At that time a small Polish Communist party emerged (PPR), and their underground fighters were known as the Peoples' Army (GL). Russian partisans became the natural partners for these Polish Communists. The Soviets, even then anticipating their future domination of Poland, began to exaggerate the sheer numbers and fighting power of Polish Communists, a practice they never abandoned.[33]

By far the largest illegal fighting force then was the hetergeneous Home Army. All references to these AK units suggest a relationship to a corresponding political segment of the Polish underground. Of the thirteen political subgroupings, nine adopted a political platform that demanded Jewish emigration after the war; four adopted a platform that advocated Jewish emancipation in liberated Poland. Indeed, Wacław Zagórski, a prominent AK figure, recorded some of the anti-Semitism he witnessed in the underground.[34] In contrast, Stefan Korboński, a Yad Vashem recipient, and the last chief of the Polish Underground State, 1939-45, firmly denies the existence of anti-Semitism in the Polish Home Army.*

However, the Home Army's specific reactions to Jews were varied and often unpredictable. Perhaps in part because of this variability the Home Army's relationship to Jews remains a highly controversial topic. Without passing judgment as to whether the Polish underground and its arm, the Home Army, extended sufficient aid to Jews, I would like to correct a mistaken impression that the only help offered came from Communists or from the Home Army's left.[35]

The case of aid offered by the Polish underground during the Warsaw ghetto uprising, itself a controversial topic, may suggest a more balanced view. On April 19, 1943, the first day of the uprising, a nonleftist Home Army unit tried to break the ghetto wall. On April 22 the same unit attacked Germans outside the ghetto and killed a few SS men.[36]

On April 20 a group of leftist fighters led by Bartoszek attacked a group of Germans forcing them to stop shooting at the ghetto. On the same day, a

---

*Personal communication, 1985.

nonleftist group from the Home Army under Józef Pszenny attacked Germans who were surrounding the area.[37] This is but a small sample of the underground's actions on behalf of the beleaguered Warsaw ghetto Jews.[38] All other helping operations also came from mixed political directions, leftist and nonleftist.[39]

Some of these efforts can be traced to organizational orders, some to individual discretions; with others the distinction between organizational and individual initiatives is unclear. It is reasonable to assume that when efforts involved help from nonleftist sources, the likelihood was greater that they were based on individual rather than organizational decisions. Finally too, the very presence of many subcenters must have contributed to the considerable discretion that rested with heads of specific AK units.

A particularly courageous underground figure, Henryk Iwański, was a leader of the Security Corps, a part of the Home Army (AK). Following his personal inclinations in aiding Jews, he minimized consultations with his underground superiors. Politically Iwański's unit belonged to the right of center.

As early as 1939 Iwański and his unit supplied a group of former Jewish officers with food and ammunition. This help continued throughout the ghetto uprising, during which Iwański lost two sons and a brother, and was himself severely wounded. None of these tragic events, however, weakened his determination to protect Jews. As late as May 6, Iwański led a group of Jews out of the flames of the Warsaw ghetto through the sewers to safety. Throughout the war he continued to save Jews, regardless of the sacrifices.[40]

A member of Iwański's unit, the rescuer Paweł Remba, was opposed to leftist politics. I interviewed him in Israel, where he settled largely because of his aversion to Communism. As a member of the Security Corps, Remba remembered that the group of former Jewish officers who were helped by Iwański formed an underground organization, ZZW. They were Revisionist and hence had a rightist political outlook. Discussion about this group was taboo in Poland after the war, because of their nonleftist ideology. The Security Corps itself was in disfavor. In fact, because he was known as a Security Corps leader, Iwański was jailed by the Communist authorities. Remba's experience emphasizes the ironic twists and turns of underground ideology and activity:

After the war I was afraid to admit to my underground participation and . . . help to the Jewish Revisionists. I knew that I could be arrested for it. Only in 1950 did I dare to establish contact with Professor Mark [Head of the Jewish Historical Institute in Poland]. I am sure that because they were inquiring about this ghetto combat group so late, much of the material was lost. Our group, the Security Corps, made three or four attempts at saving Jews and helping them fight in the ghetto. We were also taking out people and sending them into the forest. . . . Four days after the start of the Warsaw ghetto uprising we . . . dressed like Nazis and drove into the ghetto with a truck as if to confiscate furniture for the Germans. Actually this operation was arranged by Iwański and our purpose was to save Jews. We had false documents. . . . The ZZW prepared for us the people that they wanted us to smuggle out. Usually they were individuals who were unable to fight; many of them women. We saved thirty Jews this way. Our tasks were well delineated. My group brought out the staff and the people. Another group of special couriers would then take them to the [shelters].

From the ghetto we brought the furniture and the people to a store owned by a friend of mine. In it he had many cellars and hiding places. This friend was also a member of our underground and he knew what it was all about. As the Jews were unloading the furniture, some of them hid in my friend's cellar. On that particular day, as we started unloading, a courier arrived and took all the fugitives except for two women. Then the Gestapo came. Germans who lived not far away had gotten suspicious and reported some unusual activity. The Gestapo searched and searched, but found only the store owner, the two Jewish women, and me. They forced us to climb into the truck and go with them. . . . On the way our truck was attacked by an underground group of the extreme left, the Communists (GL). They did not know who we were. They threw a grenade into our car. I jumped out; the two Jewish women must have been killed. . . . I was unconscious. When I regained my consciousness I realized that my leg was wet and bloody. I dragged myself to a nearby pharmacy where . . . they took care of me. To this day I limp. You see it was the extreme left, dominated by the Communists, who forced me to terminate my help to the burning Warsaw ghetto.

Remba's account shows how subdivisions within the Polish underground could lead to tragic consequences. Also, he suggests that the postwar persecution of the AK led to a historical neglect of the nonleftist aid to Jews.

What in fact happened to the Home Army (AK) in the postwar period? How did these postwar events influence the extent of knowledge about AK's aid to Jews?

The Russians knew when they took over Poland in 1945 that the overwhelming majority of Poles, including the loosely integrated underground, were loyal to the Polish government in exile and that their support of this government was coupled with resentment of the Soviet presence in Poland. Having fought and conquered, the Russians were determined to consolidate their grip over the newly acquired territories. As a first step they set up a government made up of those whose loyalty to Russia was well tested. Communist controlled, this Polish government first undermined and eventually destroyed the legitimacy of the Polish government in exile.

Rather than acquiesce, parts of the Home Army challenged the Soviet presence by refusing to disband. Through acts of sabotage they set out to destroy the Communist influence in Poland.

Adding to the already existent chaos were clandestine activities of various underground groups not associated with the Home Army. Some of them were the splinter groups of the far right (NSZ) and Falanga.

Faced with a deteriorating situation, the Russians launched a ruthless persecution of all unauthorized underground groups. Their aim was to destroy the powerful forces of the AK. Many people arrested had nothing to do with postwar illegal operations but were targets because of former involvement with the AK. Thus some former AK members tried to conceal their wartime affiliations. In their case, to mention having shielded Jews would only stimulate interest in their other wartime activities and eventually reveal their AK affiliation.

Wojtek Kominek explained that he remained silent about his aid to Jews until 1978 because of his belief that admitting to these acts would have aroused suspicion about his other underground activities. In time his wartime memories faded and he saw no reason to bring them up. Only a newspaper article discussing righteous Christians brought him to testify before the Jewish Historical Institute in Warsaw. Kominek was very independent in all aspects of his life. He belonged to no party and identified himself only as a soldier who fought the Nazis.

Another rescuer, Jan Elewski, concealed his AK affiliation until 1950. Even then he revealed it reluctantly and only because he was afraid of being arrested. Elewski explained: *At work they were murmuring that I was an AK, they suspected me. It was then that I tried to testify that I saved Jews—to counteract the possible persecution of me as an AK.* During the war Elewski had protected seven Jews and it was this evidence that helped avert persecution. Ironically, as a Socialist, Elewski belonged to the leftist segment of the AK. But at that time Home Army affiliation overshadowed all other differences.

Adela Uszycka, a rescuer and a Yad Vashem recipient, was imprisoned twice after the war because of her continued protection of the needy, including the persecuted AK, and because of her previous AK affiliation. Actually, her political views were moderate, neither right nor left wing.

The case of the AK leader Kazimierz Moczarski points to the Communist efforts not only to punish but also to humiliate anyone opposed to them. After his arrest he was put in the same prison cell with the Nazi criminal who was responsible for the destruction of the Warsaw ghetto, General Stroop.*

Thus we know that the political climate in postwar Poland prevented nonleftist groups from publicizing their aid to Jews. It is reasonable to assume that this political climate caused some of the evidence about protection by nonleftist groups to be suppressed or lost. Keeping this limitation in mind and focusing on the individuals' political affiliations, how do the leftists and nonleftists compare with regard to rescuing Jews?

Most helpers expressed no political preference.[42] Of those helpers who did identify with political groupings, the majority were Communists and Socialists and only very few were members of the nonleftist groups and National Democrats.[43] Of the thirty-one rescuers I interviewed, about half identified themselves as leftists, mainly as Socialists.

---

*In Moczarski's case imprisonment had also some positive consequences. He wrote a book about Stroop based on material collected during his confinement.[41]

Unfortunately there is no sure way of telling to what extent results coming directly from the rescuers reflect real conditions and to what extent they are influenced by Poland's postwar political situation.

In contrast, reports from survivors on the political leanings of their rescuers might offer more accurate descriptions. Jews who mentioned political aspects of rescuing Poles had no reason to fear retaliation. They were in no way exposing themselves to threats of arrest or dangers. In short, when talking about their helpers they were just as likely to refer to those who were involved with leftist or rightist politics. To them, however, another limitation applies. Most survivors did not refer to their helpers' politics at all, hence, for the majority, such information is missing. Indeed, such indirect mention of the political leanings of helpers is available only for one-fourth of the cases. But even in this minority most rescuers are simply described as patriots.[44] Excluding the patriots from this group we are left with only thirty-one reports. In this very last group the majority were Communists and Socialists, and only a few were identified as the political right.[45] Among this tiny minority who were politically affiliated, the Communists and Socialists by far outnumbered the rest. Seen against the background of Poland's political scene, these tentative findings may gain in significance.

In Poland, before and into the war, the leftist parties were a definite minority. In the 1930 parliament the entire left wing had only 13 percent of the seats. Even if the seats captured by the center are added to this figure, it still remains a minority of 23 percent. In sharp contrast, the overwhelming majority were the government forces and the extreme right, together forming the right wing of the political spectrum. Finally, looking only at the extremes of the political ideology, the National Party captured 14 percent of the parliamentary seats compared to 1.1 percent of the Communists and the extreme left.[46] The fact that in a political setting where the left was a definite minority and yet a majority of the politically involved rescuers were leftists suggests that the left was indeed more inclined toward the protection of Jews. By the same token these findings also show that those who were identified with the political right refrained from saving Jews. Still more important, though, because rescuers who had definite political preferences were a minority, politics in general, and leftist politics in particular, could not account for participation in this risk-taking behavior.

The initial question of this chapter was, what impact did class and politics have the Poles' rescuing of Jews? Was either factor a powerful predictor for this risk-taking behavior?

Only the intellectual "class" showed a special propensity for Jewish rescue, while the peasants were least inclined toward this behavior. In contrast, among the rescuers the percentage of working and middle class was practically identical to their proportion in the population at large. What do these findings tell about class and aid to Jews? Because of the possibility of bias when considering the proportion of intellectual rescuers, I relied on the indirect reports by Jewish survivors. Yet there was still a preponderance of intellectual rescuers as compared to the population at large. In absolute terms, however, they remained a minority. Because the majority of helpers came from other classes, being an intellectual could serve only as a partial predictor for the rescuing of Jews. Similarly, even though among the rescuers, peasants were underrepresented, in absolute terms they contributed the greatest number of rescuers. Together these variable associations between class and rescue suggest that class affiliation per se was not powerful enough to account for Jewish rescue.

What about politics and rescue? Many helpers, even though opposed to the Nazis, expressed no political opinions and did not participate in illegal underground activities. On the other hand, many of those who protected Jews perceived their actions as a protest against the Nazis and hence as a political act. Compared to other motivations, however, politics played only a minor role in the rescue of Jews. Whereas political involvement in and of itself does not seem to be strongly tied to the rescuing of Jews, specific leftist inclinations are strongly associated with this risk-taking behavior. Does this mean, as some have assumed, that the political left carried the burden of Jewish rescue?

In this brief sketch of Poland's complex political situation, I suggested some reasons for the assumption that it was primarily the political left that saved Jews, and then advanced an alternative view about politics and rescue, showing that neither the politically involved nor the leftists had a monopoly on the protection of Jews.

By no means were the politically active more inclined toward helping Jews than the rest; that most rescuers were apolitical lends support to this assertion. Also, the lack of a political affiliation for most rescuers questions the assumption that most of them belonged to the political left.

Yet when those very few rescuers who were identified with specific political ideologies were considered, most of them were Communists and Socialists. That is, of the minority who were politically active, most were leftists.

More extensive participation in Jewish rescue by the left, however, should not distract us from the fact that, although a relative minority, some nonleftists also stood up for Jews. Finally, and more important, since compared to those rescuers who showed no political interests, the left is but a tiny minority, leftist inclinations cannot be responsible for Jewish protection.

Even though belonging to a certain class and espousing certain political preferences might have pushed an individual toward or away from Jewish rescue, neither factor was powerful enough to account for this risk-taking behavior.

CHAPTER 8

═══

Friendship

Because it is assumed, perhaps even seen as a given, that in times of trouble friends will do for others what strangers will not, certain assumptions about the nature of rescue during World War II gave prominence to friendship. Some have suggested friendship as the primary motivating factor, noting that most Poles who hid Jews, or helped them pass, were helping personal friends who were Jewish.[1] Others, while acknowledging the influence of friendship, have not named it the dominant motivation.[2] Still others have challenged the notion that friendship played any role at all. This last group has found that only a tiny minority of those who had Jewish friends did harbor these friends, that most others either did not consider the friendship compelling enough or, despite the pull of friendship, still turned their former friends away.[3]

My own examination of the data tends to support the following factual conclusions. Only a minority of those saved reported having been saved by friends. Many reported approaching friends first, being turned down, and then being aided by strangers. More than half simply said they were protected by strangers.[4] While Jewish disappointment at the refusals of aid by friends is not surprising, what is quite unexpected are the number of survivors who remarked that, given the danger involved in helping, they often understood when former friends denied them aid. The survivor Pola Stein, for example, articulated this position, noting: *I do not accuse anyone that did not hide or help a Jew. We cannot demand from others to sacrifice their lives. One has no right to demand such risks.*

But if the majority of Jews who were helped do not recall having been helped by personal friends, and if, as I have said, the majority of Poles who had Jewish friends did not help these friends, there is still another possibility to be considered before I dismiss friendship as a factor. Perhaps there is a correlation between friendship and help in that those Poles who did help Jewish friends also helped others who were not friends. This may be because those among the Polish population at large who had established prior

friendships with Jews were not a random group of people. In prewar Poland, it is important to remember, it was not easy for Christians and Jews
to become friends. Polish Christians lived lives very different from those of
Polish Jews, not only physically separate from one another, but often
locked into lifestyles and life patterns that did not allow for much social
contact. Those Poles who might have sought and accepted friendships with
the more assimilated Polish Jewish population still had to deal with traditionally strong Polish anti-Semitism had they tried to do so. Yet though
these obstacles made friendships less likely to occur, they did not make
them impossible.

Supporting the position that rescuers were indeed more likely than other
Poles to have had earlier Jewish friendships are the testimonies of both
rescuers and rescued. Among Polish rescuers, for example, more than half
described their having had close ties to Jews in the prewar period,[5] and a
similar proportion of the survivors reported that their Polish helpers spoke
of having had Jewish friends.[6] A look at a handful of paid helpers, on the
other hand, reveals none to have had close ties to Jews.[7] The relative
absence of friendships between paid helpers and Jews, against the greater
frequency of such friendships with rescuers, does seem to lend support to
the idea that previous Polish-Jewish friendships were more common among
future rescuers than among the population at large. But does this fact
inexorably lead to the conclusion that it was the bind of personal friendship
that led to rescue attempts? Or did the ability to maintain such cross-
cultural friendships and the decision to attempt rescue come out of similar
character traits in the rescuers?

The purpose here is not to examine whether or not most Polish friends
were willing to help, and if not, why not. I assume, without comment, that
the majority were not. Instead, my interest lies solely in what role friendship played among those Poles who did help Jews.

First, however, stepping back I look at the experiences of those who
approached friends asking for help. Their experiences illustrate the very
different responses such requests might elicit. The story of Pola Stein, for
instance, shows that when friendship was the determining factor for the
Poles, the help given could be very generous.

*When the Nazis forced us to go from the ghetto to the trains they started
to shoot at us at random and for no apparent reason. My father realized
that they were liquidating us. He took my hand and went to the side, while
all others were moving forward. The place was surrounded by villas and
gardens. We sneaked into a cellar of one of those villas. We were waiting
for all the people to move away, for the end of the deportation.*

*Just by looking at us anyone could tell that we were Jews. Our clothes
alone would give us away. This was summer and we had on the more
valuable winter clothes. When the owners of the villa found us they knew
who we were. They crossed themselves. They insisted that we leave right
away because they were afraid. They gave us water and we left. We went
directly to Jan Rybak, a peasant my father had done business with. He*

knew him, but he was not a friend of ours or anyone from whom we had a right to expect help.

Pola's father did have Polish friends, but they lived too far and he could not reach them. Much later he became seriously ill. When a trusted physician prescribed medication, neither the Steins nor their protectors had the money to buy it. The patient's condition worsened. Alarmed, they decided to turn to Mr. Stein's Polish friend who lived far away in Warsaw. Without any hesitation, the rescuer Jan volunteered to go to the city. It turned out, however, that this friend too did not have the necessary funds. But, eager to help, he sold his wife's diamond ring, the only thing of value he had, and sent the money to his sick friend. This generosity averted a possible disaster. The patient recovered.

Although the Steins were fortunate in having both the help of a friend and a mere acquaintance, Jews who had to turn to Polish friends were in no way assured of their protection.

The forms of these refusals and the reasons behind them varied, but all left long-lasting impressions on those who sought help. A refusal accompanied by a rough rejection or at times even a threat was extremely demoralizing.

Janka Hescheles, a girl of twelve, had just such an experience. Janka's mother, eager to save the child, insisted that she leave the ghetto and ask her Christian aunt for shelter. When Janka did, she met with a rude rebuff. The aunt would not even open the door for her. With difficulty, the girl managed to rejoin her mother in the ghetto. After a while mother and daughter were transferred to a nearby concentration camp. In this new setting it became much clearer that death was imminent. In desperation the mother begged Janka to take advantage of an opportunity to leave. Once more she asked her to approach the same aunt, the only Christian they knew. Remembering the terrible experience, Janka refused. The mother continued to plead: *I will take the humiliation upon myself, just go!*

This time the girl had money and instructions to offer it to the Polish woman. When Janka did, the aunt accepted her and the money. This welcome, however, was soon followed by a visit from the aunt's friend who threatened to turn her over to the Gestapo. Frightened, Janka left to rejoin her mother in the concentration camp. Much later, after her mother's death, Janka was smuggled out of the camp by members of the Polish underground. She was placed with different Polish families, none of whom she knew. Eventually, Janka ended up in a Polish orphanage where she survived the war.[8]

Much less traumatic to those who sought help were refusals that were explained by fear of consequences. Such encounters sometimes led to unusual developments. To illustrate, a group of eight Jews tried to survive in the forest and by roaming the countryside. When dangers and threats began to multiply they turned to a peasant they knew. For money he promised them shelter, but instead he denounced them. Of the eight, five perished and three escaped. In their search for shelter the three men pleaded with some Polish women they knew well to hide them, but with tears in their eyes the Poles answered: *If it were just a fine or a prison . . . but it is our lives. We also want to get through these few weeks.* The women fed them but refused to keep them.

The men returned to the forest and tried to survive as best as they could. Eventually, to protect themselves from the increasing cold, they decided to build a bunker. When the bunker was finished, the broken branches and fresh soil gave it away. Two peasant women were led straight to their hiding place. And now the unexpected happened: *Their discovery proved our blessing, they undertook to provide for us. And so they did. With unselfish devotion and solicitude.*[9] The owner of the farm on which the women lived pretended that he did not know about their existence, while the two women who had discovered them continued their selfless protection despite raids and grave dangers.

It would be incorrect to assume that all those who were helped by strangers had been refused by friends. No doubt, some Jews had no Polish friend to whom they could turn. It is also likely that some fugitives were in no position to approach such Poles. More often than not movement into the Christian world was sudden and spontaneous.* Indeed, even if a move to the Christian side was contemplated, and Christian help planned for, the actual departure from a ghetto or camp rarely happened at the designated time or in the expected way. The spontaneous and unpredictable way in which Jews reached the Christian side forced them to turn for help wherever and to whomever they could.

If Polish friends could not always be found or counted on for help, could they at least be counted on not to hurt their former friends? Here again, the answer is no. Survivor reports indicate that 12 percent either blackmailed or denounced friends who sought their aid. Because this figure comes from those who managed to survive, we cannot measure the true percentage of cases in which a request for aid elicited this response from friends. It stands to reason, however, that those who were blackmailed and denounced were more likely to have perished. No one knows their number.

My evidence illustrates that only some Jews received help from their Christian friends. Others were either denied such aid or, in the case of what appears to be a small minority of friends, were threatened with denouncements or were blackmailed. Together these findings support the assumption that friendship did not automatically lead to Jewish rescue. From the entire group of rescuers come responses that suggest similar conclusions. Ninety-five percent attributed their decision to help to compassion for Jewish suffering. Only 36 percent said that their help was motivated by the bonds of friendship. Thus neither the Jews nor the Poles perceived personal friendship as a main motivating force.

What about the second possibility—that Poles who saved Jews were at least made more receptive to the idea of helping because of past friendships with Jews. Here the testimony of rescuers and rescued seems to

---

*Recall that 84 percent of the survivors were not promised help ahead of time.

support this position. In several cases, the inability to save a friend led the rescuer into proximity with other Jewish victims of Nazi designs. Nonetheless, a closer look suggests that the influence of friendship may be more likely a manifestation of another more basic human trait, rather than a causal factor in and of itself.

As a rule Polish rescuers offered help to a variety of people: friends, acquaintances, and strangers. Of all those who extended help, 34 percent helped strangers exclusively; only 9 percent helped just friends. These figures, however, do not reflect a preference. On the contrary, if the rescuers had a choice they would have preferred to save friends. I can reasonably assume that, for a variety of reasons, they were in no position to do so.

To illustrate, until this day Jan Elewski vividly remembers his close friend whom he had helped before the war: *In my high school there were many Jews. My closest friend was a Jew. His mother and my mother were widows, and they were also close friends. Both of us were accepted at the University of Warsaw. . . . About two years before the war whenever a Jew appeared at the university, he was severely beaten. To protect my friend from this fate, I would take his student identification and get the necessary signatures. This way he did not have to come to the university except for exams.*

One evening he came to take an exam. He was supposed to join me afterwards in my room. Suddenly, without knocking, he burst into my room. Breathless, he threw himself on my bed. I could not believe my eyes. He was disheveled. His clothes were torn. Blood was dripping all over his face. A few of his teeth were missing, and one of his fingers was broken. I asked no questions. It was clear that the anti-Semitic mob had done the job.

He passed the exam. But right then he swore that he would never again set foot in this place.

Both of us belonged to a leftist student organization. He was more deeply involved in politics than I. When the war started I was recruited into the Army. He, as a known activist, went to Moscow. There he became a famous scientist and a member of the Academy of Science. In this instance Jan's friend had no need for aid, but the experience of helping his close friend seems to have affected his willingness to help others. Seven Jews, all strangers to Jan Elewski, owe their lives to his protection. While surely Jan's experiences with his high school and college friend gave him his first opportunity to help, something else seems to have kept him in a position of helping others.

Roman Sadowski, the Polish writer, and member of the Council for Aid to Jews, also worked tirelessly on behalf of strangers. He, too, got involved in rescue because he thought he might help Jewish friends.

When the Warsaw ghetto was first established, Roman took it upon himself to visit his Jewish friends: *No one knew then that the Jews were destined for extermination. At the time I thought that I could help them by* bringing them things which they were lacking. *The real help to Jews I started in 1942, when the large deportation from the Warsaw ghetto was underway. Up until then we did not realize what the Nazis had really intended. Only when the horrible extermination started, I desperately tried to establish contact with my ghetto friends. But I did not succeed. Poles just as Jews were taken by surprise. My friends perished and I could not help them. After that, except for three Jews who were only casual acquaintances, I gave help to strangers, whoever turned to me, and whomever I could find. Unfortunately, my own friends I could not help because unexpectedly they were deported and perished.*

It happened that Polish friends offered help in time but the Jews hesitated until it was too late. Others all together rejected the idea of moving to the Christian side. Until this day some rescuers are haunted by such missed opportunities and the tragic turn of events that prevented them from saving those to whom they were deeply attached.

A closeup shows how this could have happened. Felicja Zapolska was helping as a member of the underground and was also offering individualized help. She still regrets her inability to help those who were close to her: *The Polish underground had my husband take a job in the ghetto in order to help Jews. He was in charge of a factory. Because of my husband, I could send things to my friends. He tried to persuade some of them to come to our side. They were afraid of the death penalty which was so readily applied to those who did leave the ghetto. And so, while I was helping them with food, they gave me little opportunity to save them. They hesitated and were eventually caught in a deportation never to be heard from again. I must tell you about my own personal tragedy. I wanted to save someone very dear to me, but could not. This was a couple who refused my help. The woman was a Protestant like myself. She was a cousin of mine brought up in my parents' house. She was married to a Jewish man. They were very much in love. Because of him, they both stayed in the Warsaw ghetto. He was afraid to move to the Christian side and she would not leave him. When it became clear that all was lost, that the Germans were coming for them, she poisoned herself and her husband with cyanide. I loved her the same way one loves a close relative. I wanted to take them both, but because of their refusal I was completely helpless.*

Eva Anielska's painful experience is another case in point. As a member of the Council for Aid to Jews Eva never shrank from an assignment. She had saved many. Most of the time those she protected were strangers, and only rarely were they casual acquaintances. Eva had one very close Jewish friend whom she could not save: *One day I learned that my dear, dear friend was in the ghetto, which was officially sealed. I made arrangements to go in. At that point, foremost in my mind was the idea of feeding her. So I went in whenever I could and brought food. I was able to prevent her from starvation.*

Time moved on and my help to the ghetto and the Jews kept expanding, while the Nazi restrictions kept multiplying. Gradually we realized what was happening. My friend Eva had a Semitic look, red hair, and a long nose. She could easily be recognized as a Jew. Still I wanted to take her out of the ghetto. I was willing to hide her. But she refused. She did not want to endanger my life. However, during a deportation when the threat became real, she phoned me to tell me that she was in danger and asked me to save her. (Strange as it may seem, even though an action was in progress, the phone was still in operation.) I promised to save her and asked her to call me again so I could tell her how I intended to proceed. Next time I heard from her the Nazis were moving from house to house, taking out all the people. They were still many houses away from hers, but then her messages were becoming more urgent, they were away four houses, then three, then two. . . . As she talked I kept reassuring her that I would not let her down. I was becoming desperate, as I was trying to make special arrangements with a Polish doctor who promised to get me an official Red Cross car with a permit to bring Eva. He kept his promise, but when we went to get her she was not there. We arrived too late. . . . I feel that all my work did not compensate for her loss. She believed in me. She waited for me. Until this very day I have dreams about her. I just cannot forgive myself for not saving her. I wanted to write a book about her. Each time I started I would get sick.

Clearly, inability to help friends did not prevent these rescuers from risking their lives for strangers. Besides, even while they were helping friends they would also aid others. In some cases the desire to help friends started them on the path of rescue, rescue that later on, for a variety of reasons, had to be limited to strangers, or that came to include both friends and strangers. Thus, for example, initially Eva Anielska went to the ghetto to bring food to her one close friend, and only gradually became involved in a large-scale rescue operation of Jews. Roman Sadowski, too, went to visit his friends in the ghetto and later on, after they perished, became involved with the Council for Aid to Jews.

An old farmer who saved twenty-two Jews illustrates a slightly different pattern. Initially, this peasant offered shelter to one Jewish friend. Soon the friend asked permission to bring in some members of his family. Requests to include more fugitives grew, as did the number of his charges. Some of these fugitives he took in voluntarily, some only because refusal would have endangered the rest. Eventually, the good-natured Pole ended up with twenty-two individuals. All survived.[10]

Whatever the reasons, and whatever the specific circumstances, only rarely did Polish rescuers limit their aid to friends. As a rule, they simultaneously helped a variety of people;[11] some of them friends, some mere acquaintances, and some total strangers. Those who started out primarily by helping friends continued to be receptive to pleas for aid even when these pleas were from strangers.

The fact that such aid was frequently also extended to strangers, as well as the fact that a substantial proportion of Polish rescuers had no close ties to Jews, suggests that friendship cannot be a major causative factor in rescue. Instead, Polish-Jewish friendship offers at best only a partial explanation.

What could and was demanded from friends under ordinary circumstances might not have applied to the Holocaust setting. Dangerous and life threatening as the rescuing of Jews was, it simply did not qualify as behavior required from friends. The rescuing of Jews, therefore, had to be propelled by other forces, forces that went beyond the usual expectations of personal friendship.

CHAPTER 9

Religion

Back from Sunday mass, an old pious Catholic peasant, who has been hiding a group of Jews for over a year, tells his charges of the priest's insistence that Catholics bear a sacred duty to deliver Jews to the authorities. Alarmed, the fugitives ask what he intends to do. This deeply religious man first shrugs, and then with a bemused and playful smile says: *The devil finds his way even into the church.*[1]

Very different was the reaction of a Polish inmate in Birkenau, who had no trouble saying: *You Jews have crucified Christ and that is why a curse is upon you, an eternal curse.*[2]

Along similar lines is this response to a request for help from a Jew who had just run away from the concentration camp Majdanek. Instead of offering help the Pole says: *If God takes no pity on your people, how can you expect pity from a human being?*[3]

Even more telling is the story of a survivor who asks his neighbor after the war why she brought the Gestapo to his mother's hiding place. The woman unhesitatingly answers: *It was not Hitler who killed the Jews. It was God's will and Hitler was his tool. How could I stand by and be against the will of God?*[4]

Faced with Nazi extermination of Jews, devout Catholics had many choices. They could, like the old peasant, defy the urgings of a priest and save Jews. Or they could try to ignore what was happening to the Jews, rationalizing their decision by saying that it was a political rather than a church matter and that the church taught obedience to temporal authority in political matters. If approached by Jewish fugitives, they could refuse aid and ease their conscience by believing that whatever happened to the Jews was a just and deserved punishment, an expression of God's will. Employing this rationale, they could urge others to denounce Jews and, if the opportunity presented itself, to denounce Jews themselves, again using their religious obligations as justification for their actions.

What made these reactions possible? How could Christians save Jews, remain indifferent, justify, and even cause Jewish death, all in the name of God?

Throughout Poland's stormy history the Catholic Church remained powerful and strongly anti-Semitic. After the country lost its independence in 1795 the Church continued to function as the main unifying force. In 1918, when Poland regained its independence, the Catholic Church retained its prominent position as the only institution around which Poles could rally. With the emergence of the new state the powerful nationalistic Catholic Church continued on its traditional anti-Semitic path. The view of Jews as "Christ killers" and the myth that they snatched and killed Christian children to use their blood for Passover matzoh remained surprisingly intact.

Even as late as 1978 the rescuer Stach Kaminski, a Polish diplomat, pointed to the presence of religious anti-Semitism: *It's obvious when one considers that they [the Church] were, and still are, accusing Jews for the killing of Christ and for the ritual killings of Christian children. To this very day, anti-Semitic pictures are displayed in Catholic churches. For example, in Sondomierz there is a huge picture of a ritual killing by Jews. This is the sixteenth or seventeenth century Church of St. Hedwig. The picture depicts a rabbi performing the ritual killing of a Christian infant.*

Not all the teachings of the Church, however, have been as hate-provoking nor have they been as consistent. Along with anti-Semitism, the Catholic Church had demanded compassion, charity, and forgiveness for the sins of others. In fact, almost as dogmatically as the Church advocated Jewish hatred, it also required its members to reassess their own sins, to love, and to sacrifice for their fellow human beings. Ideas about love, compassion, charity, and the universal brotherhood of man were among Jesus' most important theological contributions. In a sense, then, Polish Catholics were given conflicting signals.

What happened during the Holocaust? To what extent did the Church in Poland reconcile its strong anti-Semitism with its teachings about charity, love, and compassion for all mankind? In the face of the Jewish tragedy did the Church rely on its traditional anti-Semitism or did it act in accordance with the teachings of compassion, love, and forgiveness?

Traditionally, the Polish Church relied on Rome for major policy cues, including direction for its official posture toward the Jews. Reactions toward Nazi extermination of Jews were no exception. Given this dependence, it seems likely that had the Pope, Pius XII, come out in support of the Jews, or had he condemned Nazi atrocities, Catholic ambiguities might have been resolved in favor of the persecuted Jews. This, however, did not happen. In fact the Pope's official silence, if anything, suggested tacit approval of these policies.[5] In the absence of clear directives from the Vatican, Polish clergy had also no unified policy toward the Nazi extermination of Jews.* The result was a multitude of personalized expressions.[6] At the two extremes, the

---

*The Pope did not object to the clergy's unofficial aid to Jews. He himself has been credited with personally helping Jews.

clergy and lay Catholics could lean on religious anti-Semitism or on Christian teachings of charity and universal love.[7]

Rather than focus on the official posture of the Catholic Church, here I will examine the effects religious convictions had on some Poles' willingness to save particular Jews. To what extent did these religious values reflect the official teachings of the Church and to what extent were they personal? When exploring these issues I will consider both the clergy and other devout Catholics.

All issues about Catholicism and Jews are complex and controversial, and defy easy answers. Thus the following discussion can best be viewed as a search for clues rather than an effort to establish clear-cut answers.

Faced with Jewish persecution, the Polish clergy, just as lay Catholics, could and did react in different ways. Stach Kaminski, who had a prominent position in the Council for Aid to Jews, offered a glimpse of these reactions. He recalled one incident: *We had placed a group of Jewish children in a convent. Soon we heard that all the people from this convent had been thrown out by the Germans. Despite our inquiries we could not find out what happened to them. Only after three months did one of the nuns notify me that she would like to see me. When we met I had before me a young, unattractive, mousy-looking woman, awkward, shy and timid. She told me that all the Jewish children had been moved to the new location together with the others. She added that the new convent had been inspected by a bishop. As the bishop walked among the children he said: "Here among the faces of our children I also see faces of strangers. When I come back to you they should not be here!" I interrupted her, shocked, "But sister this is outright murder. This is contrary to Christianity!" To this without fuss she said: "Calm down! Beyond the rules of the convent there are higher laws. The children will be protected." And this is how it was.*

Given this wide range of responses, how much did the Polish clergy contribute to the rescuing of Jews?

The literature offers many examples of nuns and priests who, under varied circumstances, risked their lives for Jews. It is a matter of record that many priests supplied Jewish fugitives with false documents, indispensable for survival. Others aided in other ways. For example, not far from Vilna stood a Benedictine nunnery. Seven nuns from Cracow lived there; a leader of the partisan organization, Aba Kovner, stayed with them. He was hidden by the mother superior, who had helped many others. The same mother superior also established contact between the Polish underground and the Jewish fighter organization. She wanted Poles to deliver guns to the Jews. When this could not be arranged, she personally smuggled arms into the Vilna ghetto.[8]

Jurek Wilner, a Jewish underground hero who perished in the Warsaw ghetto uprising, was protected by a Dominican convent, which also extended help to other Jews.[9]

Most of the children Eva Anielska helped were placed by her in convents and other religious institutions. As a Socialist Eva was under little compulsion to praise the Church, yet she had much praise for the clergy's support: *No convent refused me when I sent a child. Nuns helped tremendously. Mother Getter, even in 1946, refused to say how she saved Jews, because she felt that one day she may have to do it again and she should therefore not reveal the ways.* Indeed, Mother Superior Getter became a legend in her own time. Her heroism has been told and retold by many and her name stands out particularly among those who saved children.

Yet, despite these noble cases, the overall rescue record of the Polish clergy is neither consistent nor always praiseworthy. I have already described situations that show indifference and even support for Nazi policies. At times those devout Catholics who were committed to saving lives were disappointed and hurt by lack of support from others, including members of the clergy. The rescuer Genia Parska, a devout Catholic, recalled her disappointment: *When I talked about my protection of Jews, I had bad experiences. . . . I spoke about it to a prominent priest, a Jesuit. He told me: "You should not have done it, it was wrong to save Jews . . . you should have remained neutral. But to help was wrong." He must have been an anti-Semite.* Parska's experience is by no means an isolated exception. In fact, the prominent historian, Emmanuel Ringelblum, while not denying that some nuns and priests did risk their lives for Jews, looked upon their aid as mere exceptions. He felt that, on balance, the Catholic clergy in Poland was indifferent to the plight of Jews.

Ringelblum's conclusions, just as the conclusions of others who emphasize the clergy's involvement with rescue, are based on scattered case histories and casual observations.[10] For more systematic evidence I turn to my two groups: the Jewish survivors and the Polish rescuers. Here I find that only 6 percent of the survivors identified their helpers as priests and nuns. Only 8 percent of the rescuers belonged to the clergy.[11] Both in absolute and relative terms, priests and nuns did not participate extensively in Jewish rescue. In themselves these figures cast doubt on the assumption that clergy had any special propensity for Jewish rescue.[12]

Those within the clergy who did help concentrated on the rescue of Jewish children. In fact most youngsters who survived by passing stayed in convents, monasteries, and orphanages.[13] By focusing on the young, the clergy challenged a basic Nazi policy that aimed at the destruction of Jewish children. Nazi determination to eliminate the young, together with the stringent requirements for passing, made it hard to save them. Children who could talk but were not old enough to grasp what was happening, were in a particularly precarious position. Saving children meant taking drastic measures, including convincing them that they in fact were Christian. This could be achieved in part through intensive religious instructions and baptism.

Conversion of Jewish children led to accusations and counteraccusations. Admitting that the Catholic Church was active in rescuing children, Ringelblum questioned the reasons for this help. Citing the convent in Częstochowa that accepted children only under the age of six, he concluded that this age restriction meant that the Church was concerned with the saving of souls, not with the saving of Jewish lives.[14]

It is true that children who survived in convents and orphanages were baptized. It is also true that the Catholic Church welcomes converts. Still, from these two facts, it does not necessarily follow that devout Catholics saved children because they wanted converts. The basic issues still remain unresolved: Were Jewish children baptized and saved *because* the Church wanted converts? Or were these children raised as Catholics because this gave them a better chance to stay alive?

Or both?

Earlier I described the rescuer Marek Dunski, as a devout Catholic, a nationalist, and an anti-Semite. As a director of the Department of Social Welfare he succeeded in saving many Jewish children by issuing orders to orphanages and convents to accept them.

When I met Marek Dunski he presented his fellow Poles in as positive a light as possible, but was less than eager to do the same for the Jews. For example, he insisted that finding a convent or orphanage for a Jewish child was no problem, but that in taking Jewish children the Church was not interested in getting converts. While admitting that children accepted into a Catholic institution were baptized, Dunski said that this was done for safety. By becoming baptized Jewish children saw themselves as Christian and therefore could more easily adjust to their new identity.

Elaborating on this point he described a meeting with a Jewish official from the Warsaw ghetto, who questioned him about the ultimate fate of these baptized youngsters: *I tried to explain that this was the only way these children could be saved. But this did not satisfy him, for he kept insisting that they would grow up as Catholics. To this I said, "I am not a theologian but I feel that right now the main thing is to save and later on whatever the child wants it can decide to be. Maybe it will remain Christian, maybe not."*

Dunski's arguments make sense. To the young the new religion did offer safety and comfort. Most Jewish children, even those who were not in religious institutions, were deeply affected by Catholicism. Young survivors I interviewed mentioned the comforting and soothing effect that the Catholic religion had upon them. Nor were such positive appraisals limited to children whose protectors were pious Christians.

Pola Stein, whose surroundings were not particularly pious, was affected by Catholicism. In her words: *The Catholic religion gave them [the children] comfort*. In a religious setting this influence was even more intense. Leon Reik, describing the women who saved him as very pious, said: *I became religious because they were. It was a contagious disease. But this religion gave me great comfort*.

As a teenager Wacka Nowak had been protected by a university professor whose mother was deeply religious. This old woman made it her duty to instruct Wacka in Catholicism. The young girl was a willing and impressionable subject, making her new faith an integral part of her existence. On those rare occasions when Wacka forgot to say an evening prayer, sleep refused to come. She was controlled by the Catholic religion for a long time. Wacka described her attachment to her new faith: *When I met my husband, after the war, I could not have a Jewish wedding. We had to have a civil ceremony. When it came to religion I was confused for many many years.*

Children who were old enough to realize what was happening welcomed Catholicism. Those who were too young embraced it blindly. In either situation, and regardless of age, baptism and religious instructions were not necessarily forced upon the young. At least there are many cases on record showing reluctance to baptize these children without the permission of their Jewish guardians.

The survivor Malka Rosen, herself saved by nuns, succeeded in placing her Jewish-looking niece in another convent. A letter she received from the mother superior asked her permission to baptize the girl. The letter continued: *If you do not give permission it will absolutely not change my attitude to your niece. I do not fish for Jewish souls. I only am trying to save human beings.*

Ada Celka was deeply religious. During the war she sheltered the young Jewish girl, Danuta Brill. I interviewed Ada in Poland and Danuta in the United States. I heard about baptism from the Jewish woman: *My mother's father was a highly respected rabbi. Both Ada and her sister knew it. They wanted me to go to church like all the other children. It was safer this way. But to go to communion without being a Catholic would have been a sin. They thought that I should be baptized. Knowing that my grandfather was a rabbi they did not want to do it on their own. Because my grandfather was still alive, they turned to him for permission. He knew that they could baptize me without his consent. He wrote them a beautiful letter in which he gave them permission and his blessings. They baptized me only after that letter. Their priest knew that I was Jewish and he performed the ritual. He also taught me the Catechism. The Church made a deep impression on me. I was overwhelmed by my new religion. It became a spiritual escape and gave me strength.*

While baptism had made it easier for children to feel and act like Catholics, it did not always shield them from abuse; nor did it guarantee their acceptance even within the very institution that had offered them protection. Not all devout Catholics who were protecting Jewish children treated them well.

A case in point is Karla Mintz whose assimilated family had converted at the beginning of the war to Catholicism. Religious conversion, however, did not help: the Nazis murdered Karla's father precisely because he was a Jew. Karla and her mother survived the war by passing.

When Karla was five, her mother had to place her in a convent. She remembered vividly: *The nuns knew who I was. During one of her visits, I saw my mother hand an envelope to the nun which contained money. After the nun counted the money she must have demanded more, because mother took her necklace off and handed it over. I remember wondering why my mother was so stupid and why she gave her necklace away. I was angry. . . . The nuns treated me differently from the other children; they were more strict with me than with the others. Once they made me sit in a corner until two o'clock in the morning. I was a picky eater so they would right away take my food from me saying: "You don't want it, don't eat." They used to call me "cholera" [a curse word]*. Karla felt miserable and unwanted. Young as she was, when she threatened to run away, no one tried to stop her.

As a young teenager, Janina David was also placed in a convent. After a while she too was baptized. Once she developed painful abscesses on her hands and turned to a nun for assistance. Instead of the expected sympathy and relief, she was greeted with an accusation that her hands were full of scabies, a condition wrongly considered as resulting from dirt and attributed to Jews: *We do what we can to teach you cleanliness but what hope have we got against racial characteristics? What has been entered from one generation to another. . . . You people were always filthy and always will be*.[15]

Mrs. Maginski, a Polish actress, wanted to save her Jewish friend's little son. Soon, however, a few brushes with Polish blackmailers forced her to place the child in a Catholic orphanage. To the rest of the orphans it became clear that this newcomer was different. Not only was the boy mistreated by other children who called him a traitor and a dirty Jew, but the adults succeeded in instilling in him a hatred for his parents and Jews. He soon learned that Jews were bad and that they murder Christian children and use their blood for matzo. He was also urged not to return to his parents should they ever reappear. Highly impressionable, every night the boy would pray for the death of his parents. And when, after the war, his parents returned to claim him, he joined them with great reluctance. It took much effort and time to erase his anti-Semitic hate.[16]

Even if their Christian protectors mistreated them, these young converts continued to cling to their newly acquired religion. Most were unable to break their ties to their new God. Karla Mintz, for example, despite painful experiences with nuns, became totally immersed in the Catholic religion. For some time after the war, Karla had seriously considered becoming a nun. Similarly, after the war, Janina David remained a devout Catholic. Only gradually was she able to give up her new religion.

To recall, for Danuta Brill, the Jewish girl saved by Ada Celka, the Church offered "a spiritual escape." With the end of the war, Danuta stayed with her protectors, still passing as a Christian. Even when I interviewed her, after more than thirty-five years, she was bitter about her return to the Jewish religion: *I have resentment to them [the family who harbored her] because they sent me away from Poland. Had I stayed, I would have studied and I would have become somebody. . . . I did not want to leave them after the war ended. . . . When I said goodbye to my* real mother [at the beginning of the war], I knew that part of my family was in Palestine. My mother gave me her brother's address. She told me to write to my uncle when the war ended. I did not want to but they [Ada Celka and her sister] forced me to. I was afraid that I would have to leave them. They forced me eventually to leave them, because they said that they had promised it to my mother. I am sure that they had. Also they felt that my family would provide for me a better future than they could give me. . . . I resent the fact that they sent me away, it is a feeling, although I know that they suffered and had to do it. For me it was a tragedy. I loved Poland. . . . When I was leaving I felt miserable and did not care whether they were unhappy or not.

Attempts to reclaim children led to complicated and often painful reactions. Jewish parents, when faced with a reluctant and hostile child, suffered, as did adoptive parents when they had to part with these youngsters.[17]

Danuta's protector, Ada Celka, suffered as she was sending her away: *The letter from her [Danuta's] family assured me that she would live there in luxury. I explained it to the child very differently from the way my heart wanted me to . . . but because it was family and I thought that she would be better off there, I had to do it. With pain and sorrow I said goodbye to her. . . . I was a lonely person. For this child I developed strong motherly love. I suffered a lot after she left because I loved her like my own. . . . She suffered too. First, I tried to explain to her family that she wanted to stay with us, but they refused to listen. . . . I had to send her . . . it was because of duty, but against my feelings.*

Unlike Danuta, many of these young converts never returned to their own people; how many, no one knows.

From the perspective of the Jewish child, baptism and Catholicism were positive forces. Each had shielded them from danger; each had offered a feeling of security and comfort. But this positive effect still leaves unanswered the questions about the clergy's motivation.

My evidence does not tell how many of these nuns and priests helped children because of a desire to save lives and how many to gain converts. No doubt, motivations for rescue could and did vary. Moreover, some of the clergy were not fully aware of the nature of their motivation. But because the Church took no official stand toward Jewish persecution, these priests and nuns were acting in terms of their personal values. Personal inclination, in turn, might have been in part independent of the Church's desire to gain converts. Whether my reasoning is correct remains a moot question. No doubt, too, additional motivations I have not considered were also involved in Jewish rescue.

Apart from the clergy, what role did religious values play in the rescue of Jews? How many of the rescuers were compelled by religious convictions when they saved Jewish lives? My information about religious involvement is somewhat limited. Only one-third of the survivors mentioned their rescuers' religiosity, and the majority of these said their rescuers were devout

Catholics.[18] Only half of the rescuers referred to their religiosity and, of these, half described themselves as religious.[19]

What do these figures mean? Because so many rescuers were not overtly religious, it seems unlikely that religious values can account for rescue. Firsthand impressions regarding motivation support this initial conclusion. Only 40 percent of the survivors believed that their helpers were prompted by religious consideration. By contrast, recall that the overwhelming majority (81 percent) felt that Christians were helping them because they were moved by Jewish suffering. The reports offered by the Polish rescuers are even more conclusive: only 27 percent attributed their help to Jews to religious convictions. In sharp contrast, 95 percent of rescuers said they had acted out of compassion for Jewish suffering. Religious motivation thus appears to be less significant than the feelings of pity and compassion that were aroused by the Jewish persecution.

However, religion did play a part when the devoutly religious protected Jews. What was the nature and force of the religious values that guided the actions of these pious rescuers?

Ada Celka and her sister were both devout Catholics in their forties and unmarried. During the war they lost their jobs. Together with their paralyzed father, they lived in a one-room apartment and barely supported themselves by knitting. Deeply religious and patriotic, they looked upon their abject poverty as a temporary evil brought on by the Nazi occupation. Then one day their uneventful existence was disrupted by a request from their Jewish friend that they save her eight-year-old daughter. Celka remembered: *My first reaction was to refuse. After all, we had no room, father was paralyzed. People around us knew that we lived alone without children. What would we do if someone came? What could we say about a child here? But then my sister and I knelt in front of Mother Mary's picture, asking her to enlighten us. When we got up from our prayers we both said: We must take the child. Why? Because we could perish anyway. And this way maybe God would save us. We had a brother in the military who was taken prisoner by the Nazis. Maybe God would help him survive and return him to us. And we told ourselves that fate would decide what would happen to us. If we had to perish we would perish all together. The priest in whom we confided approved of our decision. He got her a birth certificate and this is how she came to stay with us.**

For pious rescuers, religious beliefs offered both the initial impetus to save and the strength to continue on this dangerous path. Sources other than my interviews also provided evidence that some pious rescuers were convinced that God looked with favor upon their good deeds and that because of His approval they would be immune to threats and dangers. At times this positive power of religion led to extraordinary deeds. One peasant built a bunker in a forest in which he hid sixty Jews. Neither denounce-

---

*Eventually, it was rumored that Danuta was Celka's illegitimate child. The priest urged them not to discourage such rumors. Such rumors offered safety.

ments nor actual searches weakened his determination to save Jews. So convinced was he that God was behind him that he had dreams in which Mother Mary urged him to continue. He saw himself as a mere instrument of God's will.[20]

Similar reasons were behind the miraculous survival of David Nassan. With a group of Jews, David was taken by the Nazis to a nearby cemetery. There they were told to undress and when they did, they were shot, one by one. When Nassan's turn came he was slightly wounded, but pretended to be dead. At night, all alone, he worked his way out from underneath the many dead bodies. Stark naked, he knocked at a farmhouse begging for admittance. This was the home of a kind, deeply religious, but destitute peasant. Except for one tattered shirt and a Sunday suit, the peasant had no other clothes. His mother-in-law, afraid of the risk, urged her son-in-law to give the Jew the Sunday suit so that he could be sent away. But the peasant would not hear of it. Instead, he persuaded his wife to keep the fugitive. He too was convinced that it was God's will. God, he argued, sent this Jew to him because he wanted him to be saved. The man survived by staying with his benefactors for twenty-seven months.[21]

As a rule, pious helpers were very independent; almost without exception they arrived at their decision alone. Determined to save human lives, they refused to budge from their position even if faced with opposition from the clergy. In continuing to protect Jews, these helpers were following their own interpretation of what God wanted them to do.

Even though Poles who continued to save Jews did not succumb to outside counterpressures, opposition from clergy and the teachings of the Church brought inner struggles and doubts to some. Influenced in part by Catholic anti-Semitism, some pious rescuers vacillated between what they perceived as the demands of the Church and clergy and of their own conscience. For those who continued to save, their personal brand of religiosity gained the upper hand.

When the Gestapo requisitioned Dr. and Mrs. Horski's garage, they had to move the Rodman family whom they were hiding there. Hela Horska approached an old retired railroad worker and farmer, Lech Sarna, a devout Catholic, who was decent, good-natured, and destitute. Lech agreed to hide the Jews in his barn. In return the Horskis supported the Sarnas and the Rodmans. The old man was convinced that Dr. Horski's standing within the community and his prominent position within the Polish underground would shield them all from danger. He also felt sorry for the Jews.

David Rodman described Lech as honest and deeply religious. Except on Sundays and holidays, Lech treated his guests with much kindness: *Essentially a highly moral and good person, he changed after each visit to Church. At such times, he would grumble, swear, and scream at us and his wife. "I am sure to lose in both worlds. They will kill me for keeping Jews and then I will lose heaven for helping Jews." He would go on and on arguing with himself, with his wife, feeling totally miserable. We tried to comfort him. He usually calmed down after a while until the next sermon.*

And so it went around and around. He was tortured, he suffered, but he did not throw us out.

Quite independently, I heard from Hela Horska that Łech regretted his decision to shelter Jews and asked her to take them back. Instead of answering him Hela called in her children and pointing to them asked: *You go to Church and communicate with God and you are willing to deprive my children of their life?* Full of doubts, suffering because of religious and moral battles, Sarna continued to protect them, and continued to vacillate between feeling good and miserable.

When Fela Steinberg's place became too exposed, she was offered shelter by an old shoemaker, Henryk Kalina. In addition to Fela, whom he knew and kept without payment, he also protected Jews for money. Fela described him as: *An uneducated but smart man, a devout Catholic who spent much of his time reading the Holy Scriptures and who reflected on the teachings of the Catholic religion. He did not care for the Jews he kept, and yet he treated them correctly. He was torn and confused by the conflicting religious values. On the one hand, he felt that maybe the Jews were being punished because they killed Christ. Some Jews he felt deserved being killed, especially those who cheated Christians. Still, there were also good Jews. He felt sorry for those. He considered himself a Christian and was aware that according to his religion one was supposed to save a soul. He was mixed up. He kept reading the Scriptures which told him that all were God's children, that all are made in God's image. He kept reading it to me trying to make sense out of what he read and out of what he saw was happening around him. He believed that had the Jews accepted Christ this would not have happened to them. That much he was sure of, but this still did not eliminate his confusion. He had a difficult time reconciling what was happening and what according to the Scriptures should have been the reaction of Christians.*

Inherent dangers in rescue, the Church's anti-Semitism, plus the absence of official opposition to the Nazi policy of Jewish extermination, all conspired to make some pious helpers doubt their actions. Those who withstood these pressures and continued to save lives had relied on their personal convictions, their own interpretation of religion.

In summary, I have bypassed the important but controversial issue of the Catholic Church's official posture toward the Holocaust. Only briefly did I note that the Vatican and the Polish Church had failed to officially condemn the Nazi extermination of Jews. To this very day, the Vatican continues to be silent on these important questions, leaving its wartime position unexplained.

The silence of the official Church, however, did not prevent some Polish clergy from risking their lives for Jews. Compared to other groups, their overall contribution to Jewish rescue seems modest. And those priests and nuns who did offer help to Jews tended to focus on children. These children were instructed in the Catholic faith and baptized; conversion of the young diminished the danger of discovery. With younger children in particular, baptism convinced them that they had become full-fledged Catholics. This, in turn, reduced the possibility that they would unwittingly reveal their origin. In addition to greater safety, the Catholic religion offered these children solace and a feeling of security.

Such conversions, however, led to the accusation that the clergy was motivated by the desire to convert rather than by the desire to rescue.

Motivations are rarely, however, directly deducible from actions. The possible motivation of pious rescuers to convert Jews does not nullify the fact that they were saving lives. To achieve a more balanced view, I have shown that at times both the clergy and lay Catholics were reluctant to use the danger of the time to justify baptism.

Regardless of the clergy's motivation for saving Jewish children, two important but often neglected facts remain. First, clergy and lay Catholics were no doubt prompted to save lives by a variety of motivations, not all of which were conscious. The possibility that some of these motivations grew out of a desire to convert ought not to detract from a basic and more important consideration: that pious rescuers, like all others, were risking their lives for no tangible rewards. Indeed, it is the selfless, self-sacrificing, and lifesaving aspects of their actions that deserve special attention and not the questionable secondary gains that might have resulted from the acquisition of new converts.

Aside from these special motivations, what effect did Catholic beliefs have on Jewish rescue? To devout Catholics, religion offered a range of possible reactions. Those who interpreted their religion in concrete terms and saw the Jews as "Christ killers" could interpret the suffering of the Jews as a sign of God's will to punish them for their past sins. Others who read the Bible could see in a Jew someone who deserved help and compassion. In fact, in the name of religion some Polish Catholics protected Jews, others remained indifferent to Jewish suffering, and others denounced Jews. If religious convictions and involvement could lead to so many different interpretations and result in so many different actions, they cannot be seen as responsible for the rescuing of Jews. Doubt about the explanatory power of religious values is compounded by the fact that most rescuers did not see religion as occupying a central role in their activity. In short, Catholic religion and values do not qualify as complete explanations for Jewish rescue.

What about the rescuers who were devout Catholics? Given the ambiguous messages of the Catholic Church plus the Vatican's refusal to take a stand on the Nazi annihilation of Jews, pious rescuers had to rely on personal rather than on official religious values. It is certain that religion played a positive and significant role in the protection of Jews by pious rescuers, but these helpers seemed to be religious in a special way. They were independent in their interpretation of religious values, and this independence prevented them from blindly following the teachings of the Church. Moreover, their compassion and strong moral convictions, though expressed in terms of religious values, seemed stronger than concrete teachings and images of the Church. Therefore, it seems reasonable to conclude that it was these moral convictions rather than religion per se that made them rescue Jews even when faced with opposition from the clergy. In short, religious involvement—backed by an independent spirit and a firm sense of morality—was apparently the only kind of religion that could have been instrumental in the rescuing of Jews.

CHAPTER 10

A New Theory of Rescue and Rescuers

Thus far I have examined many influences on the decision to undertake selfless rescue—social class, political beliefs, degree of anti-Semitism, extent of religious commitment, the prospects of monetary reward and friendship. While each of these may offer a partial explanation, none is a fully reliable predictor of precisely who would attempt the protection of Jews. If the actions of righteous people cannot be explained by reference to readily definable indicators, perhaps the problem is that I need to go a level deeper. Perhaps I ought to search for more basic core characteristics.

In addition to my own examination, thus far the literature contains a few isolated efforts of others to study systematically the characteristics and motivations of the righteous. One notable case is Perry London's research. Published as a single exploratory paper,[1] it offers a number of intriguing observations that have served as a starting point for others who have studied the topic and who, like London, view Jewish rescue as a form of altruism.[2]

If the selfless protection of Jews suggests a form of altruism, perhaps it is valuable that I stop a minute and examine certain available studies in this area.

Though vast, full of fascinating formulations and possibilities, the literature on altruism contains both contradictory and unresolved issues. Some question the very existence of altruism. Anna Freud, for example, reasons that because givers enjoy the act of giving, the satisfaction they derive from helping others belies the selflessness of their acts.[3]

A less extreme position is taken by those who assume that a norm of reciprocity governs all human behavior. The implication here is that the giver always receives something in return.[4] In a strict sense this, too, questions the existence of altruism. Others do not deny the existence of altruism but insist that expressions of this kind of behavior are extremely rare.[5]

Closely related to an understanding of these issues then is the question of defining altruism. Often interchangeably referred to as aiding, helping, and prosocial behavior, altruism, in general, refers to doing things for others without expectation of external rewards. Only very rarely included in discussions of altruism is the possibility of sacrifice or cost to the giver. Interestingly, in the biological sciences "altruism is ordinarily defined as self-destructive behavior performed for the benefit of others."[6] To the extent that Christians who saved Jews were knowingly risking their lives, their actions did contain this self-destructive element.

For other reasons as well, the literature on altruism must be employed carefully, for there are important distinctions between the data on which this field relies and the rescuing of Jews during the Holocaust. As a rule, most altruism studies focus on single acts of help. Probably the largest part of *this* kind of research deals with bystander intervention. Many of these bystander studies explore the conditions under which people will or will not plunge into a dangerous situation to protect others.[7]

While some righteous Christians may have begun their protection of Jews in a spontaneous and unpremeditated way, their subsequent actions were not limited to isolated encounters. In fact, those who confined their aid to a single act do not fit into my definition of rescuer. Thus, those who studied help in an experimental situation were concerned with single acts, whereas I focused on continuous aid.[8]

Another more basic difference concerns the very nature of the activities. Aid to Jews during the Holocaust held out the possibility that the act would make necessary the ultimate sacrifice by the giver—the loss of his or her own life. This situation cannot be experimentally reproduced.[9]

Third, most researchers in this field have hesitated to establish theoretical constructs and have chosen instead to work with operational definitions.[10] To illustrate, such operational definitions have included picking up unmailed letters and mailing them,[11] dimming one's headlights for an approaching car,[12] writing letters on behalf of others,[13] picking up someone's dropped groceries,[14] and giving money when approached by a stranger in a supermarket.[15] Clearly, these and similar helping acts bear no resemblance to the rescuing of Jews because the price to be paid is insignificant when compared with the risk of one's own life. The prospects of these studies enlightening us about the self-sacrificing behavior of righteous Christians is doubtful.

Finally, too, the complex field of altruism has produced no uniform set of theoretical explanations.[16] For many reasons, then, insights gained by those studying altruism can serve only as limited guides to understanding righteous aid.

Thus, if we view these rescuers as altruists, bypassing other definitions of altruism, we can define their behavior as that which is carried out to benefit another—with a possibility of very high, rather than inconsequential, personal costs to the giver.[17] When dealing with the rescue of Jews a distinction between two types of altruism seems appropriate: normative and autonomous.

Normative altruism refers to helping behavior demanded, supported, and rewarded by society. In contrast, autonomous altruism refers to selfless help, which is neither reinforced nor otherwise rewarded by society. Indeed, autonomous altruism may be opposed by society and may at times involve grave risks not only of physical injury but of social ostracism.

Thus, society establishes the standard that a mother should donate a kidney to her child, that a child should aid an ailing parent. However, social acceptance does not require of its members that they donate their organs to strangers much less so that they risk their lives for strangers, particularly not for those whom society despises. Indeed, saving Jews in Nazi-occupied Poland put the actor in conflict with his society's expected values. In this sense, then, those who without regard for external rewards risked their lives to protect Jews clearly fit the definition of autonomous altruists.[18]

What kind of person was it who could withstand all the pressures and who, despite all the threats and dangers, stood up for the persecuted Jews? What prompted these people to risk their lives in an undertaking that showed little promise of community support?

Referring to rescuers as altruists, London addressed himself to some of these issues by interviewing 27 helpers and 42 Jewish survivors who benefited from selfless aid. While all his interviews took place in the United States, they were conducted with people who came from many European countries.

Staying close to the data and thus employing the inductive method, London first notes that the rescuers came from a variety of backgrounds and thus represent a socially heterogeneous group. Stressing this social heterogeneity, he found that they still had three characteristics in common: (1) a sense of adventure; (2) a strong identification with a parental model of moral conduct; and (3) a sense of being socially marginal.

Marginality in London's terms refers to standing out within one's environment, not being fully integrated within it. I think that adventuresome behavior as described by London is but a form, an expression of marginality; hence the two characteristics, adventurous behavior and marginality, may be seen as shades of the same characteristic.[19]

Suggested by London's observations are two interim conclusions: First, that helpers were aware of their marginality, and second, that the rescuers' need and desire to save others can be traced to familial values.

London's idea of marginality is closely related to social integration. Because I was sensitive from the start to the ways in which the rescuers related to their environments, I was influenced by his concept of social marginality. But especially after I realized that class, politics, religion, anti-Semitism, friendship, and money each offered only a partial explanation of rescue, I examined the many and varied experiences of helpers with a renewed openness. Without a systematic theory to test, I knew that I would have to stay close to the empirical data, relying extensively on the inductive method. Guided by the concept of altruism and social integration for some preliminary clues, I looked at scattered cases of rescuers from different countries. Already this step yielded some intriguing though tentative conclusions.

Without exception and regardless of the country they came from, helpers insisted that for them saving Jews was a natural duty. In the overwhelming majority of cases, their protection of Jews fit into an already established pattern of helping the needy. These righteous took their obligation to provide help for granted, finding it hard to explain. One of their typical answers was: *Of course I helped them, but I cannot tell you why.*[20]

Additional evidence shows an inability to blend into their environment, a condition identified by London as marginality.

To illustrate, Jeffrey Jaffre, a Christian helper, had a Swiss father and a Latvian mother and later on a Jewish stepfather. He seemed to be suspended between not only different cultures, but different countries and religions. As a young student, he had acquired a reputation as one who would defend the persecuted and the helpless. In 1940 he helped save 50 Jews and he helped many others also. Yet, he categorically denied that his actions were heroic or special.[21]

An incident from Belgium offers a glimpse of another similar case. A Belgian Jew recalled his search for someone who would take in his young son. He approached a number of upright decent citizens whom he knew well. He met with continual refusals; none was willing to take the child. Time was running out. On the verge of desperation, the father made one final attempt and asked a French woman, whom he barely knew. She was unmarried, and for years had had an open love affair with a married doctor in the small town. She had the reputation of a rebel and was said to have associated mainly with bohemians. Unlike the conformist and law-abiding citizens who refused the father's pleas, she agreed in a matter-of-fact way to take the boy and did indeed save him.[22]

We find the same pattern in a Huguenot community in France, known for its heroic protection of Jews. In Catholic France the Huguenots are distinct from the rest of the population. André Trocme, a pastor who was responsible for making possible massive aid to Jews, was particularly individualistic. Half German, half French, he was atypical within this atypical community. Those who helped him in these efforts to save Jews also showed expressions and degrees of individuality and their aid to Jews was a continuation of past acts of charity.

Typically, these villagers perceived their protection of Jews as natural and denied that their actions were extraordinary or heroic. Finally, too, they insisted that it was immaterial whether those they helped were Jewish or not. All that mattered was that those on the receiving end were suffering and helpless.[23]

At once very similar and different is the case of the Dutch underground figure Joop Westerweel.[24] Joop was a Christian anarchist, a pacifist, a teacher, and a fighter for social causes. As a young man he had emigrated to the Dutch West Indies where he incurred the displeasure of the authorities because he spoke out against the exploitation of the natives.

After six years Joop returned to Holland. At that point he established a school that relied on the Montessori principles.[25] During the war, Westerweel's behavior was also special. He and his wife Will placed their small children in foster homes, gave up all previous employment, and joined the underground, where they became totally absorbed in fighting the injustices committed against Jews. The couple organized a network of hideouts and smuggled Jews across the border into France and then into Spain. After a year Will Westerweel was arrested, tortured, and sent to jail and then to a concentration camp where she survived fifteen months of confinement. Her husband was not as fortunate.

Aware that the Gestapo knew about him and was looking for him, Joop kept changing documents. In the end, the Nazis caught up with him. He was arrested in the summer of 1944 as he tried to smuggle out of the Edinhofen camp two Jewish girls. A false identity card failed to protect him. The Nazis soon discovered who their prisoner was. They sent him to the camp of Vught. Five months of torture followed. The underground's continuous efforts to free him were frustrated. Only after five months, with the help of a Dutch camp physician, did they conceive of a definite rescue plan. But this too failed. The doctor was caught with the plans of Joop's escape and paid with his life.

When he was being tortured, Joop Westerweel worried about others. A message smuggled out to his friends read: *You know I shall never betray you. . . . I am a very ordinary person so please don't idealize me. . . . I have just to pass these difficult days. If my fate is doomed, I shall go as a man.* True to his words, he never revealed a name. Instead, as Joop Westerweel was led to his death, he recited a freedom song he loved.[26]

How did the rescuers in my own sample compare when examined at close range? I found a cluster of shared characteristics and conditions similar to some of those found among the scattered cases from different European countries. They include: (1) The inability of the rescuer to blend with the environment, a characteristic that resembles closely London's concept of marginality. But because marginality has a negative connotation, I have chosen instead to employ the word "individuality" or "separateness" to describe this characteristic. (2) A high level of independence and self-reliance that causes these individuals to pursue personal goals regardless of how these goals are viewed by others. (3) An enduring, strong commitment to help the needy that began before the war and that included a wide range of activities. (4) A matter-of-fact attitude toward rescue that sees it as a mere duty, which explains the repeated denials of rescuers that their protection of Jews was extraordinary or heroic. (5) An unplanned beginning to rescue efforts. (6) A universalistic perception of the needy; the ability to disregard and set aside all attributes of the needy except their dependence and helplessness.

These basic traits and conditions cut across the conventional ways of classifying people: class, politics, religion, friendship, anti-Semitism, and money motivation. Closely interdependent, both separately and together, these characteristics and conditions demand careful attention.

Starting with the rescuers' separateness, in what ways did it manifest itself? One of the rescuers, Felicja Zapolska, a journalist-lawyer, came from an upper-class family. Many of her aristocratic ancestors had distinguished themselves in Poland's fight for independence; a fact she kept emphasizing again and again.

Throughout the war Felicja tirelessly helped Jews by escorting them from the ghetto, finding them shelter, and providing them with food and false documents. For two and a half years, she hid a Semitic-looking woman in her own apartment. On the surface there was nothing very unusual about Felicja Zapolska. She also told me that her family included a Jewish great-grandmother: *I never knew my great-grandmother, but I heard about her from my grandmother, who was proud of her mother and wanted me and her children to remember her Jewishness. I loved my grandmother, and this made an impression on me. Then a terrible thing happened in my family: my mother married a German industrialist. Because of him I was Protestant. I had a German name, my father's name was Adolf. . . . My parents separated when I was twelve. My father went to Germany and I saw him only occasionally. My mother's family was very patriotic, very Polish. I too was like them. At the same time I was brought up with a great deal of respect for the Germans. I was taught that they were a wonderful nation and a well-organized, well-ordered society. During the war the Protestant community split into two parts: Polish and German. I was pressured to move to the German side. I was very upset. I eventually moved to a different Calvinist church in which there were no Germans. Perhaps my strong desire to oppose the Germans was somehow more violent because of my German blood.*

Unlike Felicja, Olena was definitely at the bottom of the social ladder. An unmarried old woman, a poor and landless peasant, all her life she was ridiculed and called a fool by the rest of the villagers. During the war, Olena, protected a Jewish girl on whose behalf she faced many dangers but for whom she was willing to sacrifice all. As she confided to her Jewish charge, she was aware of her peculiar position: *These people [all the villagers] were not interested in me. I did not belong to them. I had always been an outsider. I never had anyone who liked me. . . . At least I have the consolation of knowing that I was protecting an innocent helpless person.*[27]

The Lasows were the Polish couple who aided Rina Ratner. Their aid, however, did not end with the Ratners. Tadek also worked as a laborer in the concentration camp Majdanek. Each day, Tadek would give his lunch to the starving Jewish inmates. He also tied tobacco leaves around his body and smuggled them into the camp. Even when the tobacco left sores on his skin, he continued this practice, arguing that the joy his gift brought to the unfortunate prisoners by far outweighed his own discomfort.

Who were the Lasows? On the surface they seemed like typical working class Poles. Tadek, the head of the household, was an unskilled laborer who had difficulty supporting his family of five. The family's meager diet consisted mainly of cabbage and potatoes. But Tadek was also a Communist. He knew well that if the Nazis discovered his political affiliation, he and his family would not escape punishment. Earlier, because of his Communist activities, his father had been sent by the Russian czar to Siberia.

The Lasows were outsiders in one other way. They refused to have their children baptized, and none of them ever attended religious services. Living in an area where people took their religion seriously, they were social outcasts because of this rejection of religion.

And in still other ways, the Lasows were different. Edna came originally from Germany, and at a time when this fact could have been advantageously used, she chose to reject all German connections. In addition, the entire family was, on principle, opposed to anti-Semitism and did not hesitate to make their views known. Rina was aware of their unique qualities: *They simply saw people as good or bad. They were not a part of their surroundings, they were very different. This was obvious to anyone. When on special holidays the priest came to the neighborhood to bless each house, they were the only ones who refused him entry. They had nothing to do with their neighbors. They were definite outsiders, marvelous people who never became a part of their surroundings.*

On the surface the peasant Maria Baluszko seemed to have no distinguishing features. She saved five Jews and helped many others. She never attended school, but in her surroundings this was not unusual. What was unusual, however, was that she belonged to the richest family in the village, a family that employed a number of servants. Maria was deeply religious and her close friendship with the priest gave her an added distinction. She and her family were distinguished from the rest of the peasants in still another way. Before the war she defended Jews against anti-Semitic attacks. In her village this was unusual and strange behavior that could destroy a person's reputation. Baluszko readily explained: *The anti-Semites used to say that the Jews do not work physically, that they use people. So I told them you should also try easy work. When your head is no good for mental work, you work physically. Then they said that I stand up for the Jews. That I am a Jew lover. This is what they said to my face and behind my back. But I was not afraid of them.*

The Baluszkos were different in other ways as well. Unlike most families in the village, they were very peaceful; beating, drinking, and quarrels, so much a part of the peasants' lives, were unknown in their household.

Last we come to the testimony of the survivor Wizling, who describes how a known thief, a landless peasant who lived in utter poverty, a man despised and feared by the rest of the community, rescued him. Wizling describes this man as a "marginal" character, avoided by all, a true outcast. Still, during an "Aktion"[28] this man dared to remove and hide twenty-six

Jews who were on their way to destruction. The same Pole, despite continuous threats and dangers, sheltered and saved the Wizling family without wanting or expecting anything in return.[29]

As interesting, in his testimony, Wizling adds a commentary, noting that those Poles who were on the periphery of their communities were more likely to save Jews than were those who were well integrated into their social surroundings.

Independent of their social standing, these rescuers seemed, to one degree or another, not fully a part of their own communities. What comes with separateness? As I continued to search for factors associated with this quality, it became clear that whatever else separateness meant, it was unrelated to the degree of prestige a person enjoyed within a community. Moreover, many of those who stood out as clearly different by their own descriptions or by the descriptions of others claimed not even to be aware of their separateness. If this is so, it challenges London's assertion that the rescuers were generally conscious of their individuality. Moreover, as the forthcoming cases show, this individuality could vary in many intricate ways.

One of those who clearly knew that she was different was Adela Uszycka, a Yad Vashem recipient, whom I interviewed in Poland. She married young but her marriage lasted only a few months. Her husband left her abruptly. She never remarried, lived alone, and supported herself, not very adequately, as a masseuse. Adela was aware of her peculiar position in society: *I was always an individualist, I had my special circle which was made up of theosophists. I don't eat meat, I believe in the unity of life. I am different from other people, but this does not bother me.*

In some cases, even though these Poles did not perceive themselves as outsiders, the information they offered put them easily into this category. Thus, Franek Dworski's inability to fit into his environment was easily apparent and required no special sources of information. His position as an outsider was of long duration. His World War I injuries had placed him into the official category of a war invalid. His own physical handicaps spurred him to fight on behalf of others who were helpless. Soon he became the secretary of the association of invalids. In this capacity he initiated strikes and organized demonstrations, all of which prevented him from finding regular employment. His fight on behalf of invalids' rights eventually led to trouble with the authorities. To avoid persecution he was forced to run away from Warsaw. Still, Dworski never stopped standing up for unpopular causes and found himself frequently in opposition to existing authority. As a Socialist, he objected to much that was happening in prewar Poland and did not hesitate to voice his objections.

World War II also gave Dworski yet another opportunity to fight injustice. As a Socialist underground figure, he spared no effort on behalf of the persecuted. He was and continued to be an outsider and rebel, but did not see himself in these terms.

In contrast, Ada Celka's case was less obvious. As a governess, Celka had a limited income. During the war her modest earnings dwindled further, but this did not prevent her from keeping and supporting the Jewish girl Danuta Brill. When I interviewed Ada in Warsaw she gave me few clues about her social setting. Here and there vague hints emerged, none of which added up to a clear picture. True, she never married, but there were valid reasons for that. She and her unmarried sister were committed to taking care of a paralyzed father. Deeply religious, Ada assured me that God gave her the strength to protect the helpless. This too was not unusual. Still, her command of the Polish language was impressive and far outstripped her social position. But this too did not add up to separateness. More direct information came to me through the Jewish woman Celka saved. Unhesitatingly Danuta Brill described Ada as *very unusual, highly principled, and highly educated. She did not fit into her environment at all, which was made up of common laborers. She was very different. She was highly respected by her neighbors and looked up to. People accepted her as something better and looked up to her, but she was an outsider. She was never one of the people. . . . She was superior to the rest, and was treated accordingly. . . . She herself had a feeling of class.*

Most Jewish survivors, including Danuta Brill, were unaware of their helpers' separateness and individuality. Only after I had raised the issue did they realize that their helpers were different and that they did not quite fit into their surroundings. Having made this observation, most of the survivors admitted that they had never thought along these lines, but found such new insights provocative.

Thus, the survivor David Eckstein became intrigued by such questions. When commenting about his helpers he agreed that: *They did not fit into their surroundings. Such Poles were not typical but very different. . . . I never thought about this problem before. As far as I remember they had no social contacts with others. This must indicate their relation to their own group. They had no friends. They were definitely different. I believe that I was too young then to realize. Now I am convinced that they were different, but I cannot tell you exactly how.*

Leon Reik was clearer about his protectors' peculiarities. He described to me one family with whom he stayed for over a year: *I was placed with a strange family. They were all women, four sisters and a niece. Three of the sisters had never married. The one who did marry and who had a daughter was at that time dead. To me they all looked old, very old. They did not always stay together: for much of the time two of the sisters and the niece lived in Warsaw. I and one of them stayed permanently in the village on a big farm close to Warsaw. All four sisters were teachers. The one who ran the farm and who stayed with me gave up teaching. She was more than strange, she was weird. She never washed, never! And when the others came to visit they too did not wash. Even the niece who was much younger fell in with their pattern and stayed away from water. In a small village like this, and because they were teachers, they were socially above every-*

one else. They, in fact, felt superior to all others and refused to have anything to do with them. They had contact only with the priest. Otherwise they were completely isolated.

They must have had more than twelve cats. They could not drown a cat, nor could they turn away a stray dog. So the place was swarming with strange animals, some sick and some healthy, all of which were fed and taken care of. They were very religious. They told me that Jews killed Christ. This convinced me that Jews were not as good as others. Because of them I became very religious and felt guilty because I was a Jew.

Vera Ellman's protectors were intellectuals. Mr. Lofar was a director of a theatre, his wife a schoolteacher. Vera knew Mrs. Lofer through their past work in the local school. They liked each other but they were only casual friends. When in a strange town, penniless and homeless, Vera turned to the Lofars for help, and they received her with open arms, giving her their son's room and insisting that she stay with them free of charge, which she did until the end of the war. The Lófars were very religious, very patriotic, and involved in the Polish underground. Both liberal and tolerant, they were not contaminated by the prevailing anti-Semitism. When asked how the Lofars related to their surroundings, Vera showed surprise. She had never thought about them in these terms, and she was formulating her thoughts as she was answering me: *Within their environment, they stood out. Somehow they were not fully integrated into it. The mother of Mr. Lofar was a Hungarian. Mrs. Lofar's sister married an Austrian, and her son was an SS. When the SS man came to Poland he tried to see his aunt, but she refused to meet him. Later on he was killed in action. It could be that because of the German connection, Mrs. Lofar was particularly sensitive about her nationality. Both Lofars had relatives in Czechoslovakia; both were more cosmopolitan than others around them, and in general seemed to have had more tolerant views about life than anyone else I knew. Who knows, maybe the fact that they were so different somehow explains their help.*

Many more examples could be offered. Indeed, of the thirty-one Christians I interviewed, all but two can clearly be classified as individualists. The exceptions are Hela Horska and Marek Dunski, both overt anti-Semites, nationalistic, and deeply religious. Their aid to Jews was motivated by religious values that required them to repent for the harm that they thought their anti-Semitism might have done to the Jews. I do not believe that these two cases deny the existence or the close association between individuality and rescue, but only that they represent interesting exceptions, if not simply variations.

That the rescuers did not blend well within their environment is also evident when I compare what survivors said about their selfless helpers and about their paid helpers. Those Jews I interviewed directly all described their rescuers as special people. When it came to paid helpers, however, their reactions were very different. Of the ten paid helpers mentioned, only two were described as individualists.[30] Genia Huber, then a teenager, described her Polish helper as a woman from an aristocratic but impoverished family. She had inherited large debts at her husband's death. The money she received from Genia's father kept her from starvation. Genia referred to her as an intellectual who was isolated from everyone else except her family. Asked how this paid helper fit into her surroundings, Genia's unhesitating answer was: *She did not blend into the environment. I would say she was an oddball. Different. I wonder if there was not a Jew in her family tree somewhere?*

The other paid helper described as an outsider was the peasant who protected Lola Freud and her brother. A bachelor, this Pole lived alone and had no contact with the other peasants. Lola referred to him as someone who: *. . . felt himself above the community. He felt himself more intelligent than the community. He wrote Polish well, was interested in politics. He was a patriot who hated the Germans. Most of the others were farmers and really did not care about anything else except the church and farm.*

Both direct and indirect descriptions indicate that 72 percent of the entire sample of Polish rescuers stood out sufficiently to be defined as individualists.

Results from the larger sample of Jewish survivors are less conclusive, because this kind of information is often unavailable.[31] Absence of data, however, does not in and of itself challenge the basic finding. Enough available evidence does show that these courageous Poles were at best only loosely integrated into their communities.

Individuality or separateness appears under different guises and, no doubt, varies in more ways than shown here. Moreover, its effectiveness and impact on rescue depends on and is related to other attributes and motivations. What does being an individualist entail, and what else does it suggest?

Being on a periphery of a community means being less affected by the existing social controls. With individuality, then, come fewer social constraints and more freedom. More freedom in turn means more independence. A high level of independence has other important implications. Indeed, freedom from social constraints and a high level of independence offer the opportunity to act in accordance with personal values and moral precepts, even if these are in opposition to societal expectations. In short, if rescuers are less controlled by their environment and more independent, they are more likely guided by their own moral imperatives, whether or not these imperatives conform to societal demands.

Are these rescuers in fact self-reliant? And to what extent is their independence expressed by their ability to follow personal values even if these are not approved by society?

Once again for answers I turn to the actual cases of helpers. To escape Nazi persecution, the Polish officer Jan Elewski settled in a remote village. In this new place he established underground connections and worked officially in the mayor's office. Because Jan had attended the Warsaw University, his boss, the mayor, offered him more responsibility than normally associated with the position of administrative assistant. The uneducated peasants, one of whom was the mayor, appreciated and respected Jan, but never accepted him as one of their own. Instead, they continued to treat him with guarded suspicion.

By 1943 the Nazi law requiring Christians to deliver Jewish fugitives was well established. In this remote village people preferred not to act on it, even though many knew that Jews were hiding in the surrounding area.

When, however, the whereabouts of these fugitives became an open secret, the Poles became fearful lest the Nazis punish them for failure to report. They decided to act. One day, an official delegation appeared in Jan's office and, in front of witnesses, publicly announced that seven Jews were hiding in a nearby field. This happened on a day when the mayor, one of their own, was away and Jan was in charge. Listening to them, Jan understood that they preferred to burden an outsider with the unpleasant task of delivering victims to their death. By denouncing the Jews these Poles had performed a duty demanded by the Nazi law. Now they knew that it was up to the mayor's representative to do something about it. Elewski also knew that for him noncompliance could result in his execution. Therefore, as a village official Jan went through the motions and followed the rules. He wrote down the report and promised to notify the authorities. But he could not act upon his promise. Instead, he warned the Jews about the impending danger, moved them to another hiding place, and supplied them with food until their liberation.

Does his level of independence, life philosophy, ideas about right and wrong, give any clues to his actions? Let him explain: *I never worried about the opinion of others. I tried to stay away from too many people. . . . I did what I had to do. . . . My family would have disowned me had they thought that I was not helping people who were being destroyed. It was my duty to do it, so I did it.* Asked about the differences between those who did or did not help, Elewski felt that those who did not help *were self-centered; they thought only about themselves. Those who helped had a strong moral base, otherwise they don't differ very much. They must have been guided by some ideology that made them do it, by a feeling of duty. After all one had to be at peace with oneself.*

The close connection between freedom from social constraints, independence, and the effectiveness of personal values is illustrated by Emil Jablonski, who for two years kept a Jewish friend in his one-room apartment. *To be sure I am influenced by the opinions of others but only up to a point . . . regardless of how others feel I let myself be guided by my own personal moral values. I am an individualist, a cat who moves on his own road. I always have a special and individual approach to things, my own which is often different from the way others do. For me moral values are of the greatest importance.* And gentle Ada Celka became adamant and emphatic as she said: *No, I do not depend on the opinion of others. I try to do what I consider to be right and decent. I tried to live in accordance with my religious and moral precepts. If someone has a good or poor opinion of me or my actions this has no effect on me. I have to act in ways which I consider to be right and it does not matter to me how others look upon it.*

Janka Polanska, a Yad Vashem recipient, saved ten Jews by hiding them in her apartment. Does she depend on what others think of her? *I have to be at peace with myself, what others think about me is not important. It is my own conscience that I must please and not the opinion of others. My father always told me that I should act in a way which should not embarrass me. . . . Public opinion is fickle, it depends on the way the wind blows. . . . At one point during the war I had to decide whether to follow my conscience or save myself. One day, Adaś, the Jewish man who stayed with me, left and did not return. The Polish police came to search the house. We had a good hiding place and they did not find the others. They told me that Adaś was at the police station. I was advised by all who knew about it to run away. The situation appeared hopeless. To try and save him would be suicidal. Adaś looked very Jewish. Also, as a man his identity could be easily checked. But I could not follow the advice of others. Instead, I went to the police station. They wanted a large sum of money which I managed to get. They also demanded that I give them a grand piano and some valuable paintings. I was ready to give them everything. After I met all their demands, we were afraid that they would continue to blackmail us. They knew where we lived and they knew at least about the existence of one Jew. I had to change apartments. I decided to tell the AK (Home Army) that I needed an apartment. I told them that if they could get me one I would keep arms for them. They did. All of us moved to it safely. The underground never knew that I was hiding Jews. And so from then on, I had two illegal things, arms and Jews.*

When I got Adaś from the clutches of the police, I felt so light it was as if I had wings. I felt so gratified; I knew I acted correctly and that I had no reason to be ashamed of myself.

Staniewska, now a physician in Warsaw, protected a Jewish girl during the war. Compared to others, is she more or less concerned about people's views? *I think that I pay too little attention to what others think. I speak my mind frankly, I say things that others would not say, when others would be more cautious. I am an individualist. I rather like to be in opposition. I never accept things unless I have a chance to weigh them.* And what does Staniewska have to say about the difference between Poles who did and did not help Jews? *Morally they are different. Those that did not help are self-centered, selfish; those that helped have a stronger moral base.*

Franek Dworski's past attests to his independent spirit. As I mentioned earlier, he fought for the rights of the veterans and, because of it, had to run away to save himself from persecution. During the war he was helping all who asked for help. While Franek did not dwell on his courageous actions, he was conscious of his independent spirit. *I always had my own goals and aims in life regardless of how others felt about it. I fought against what I considered to be wrong. . . . This did not mean that others agreed with me but I was not bothered by it.*

Maria Baluszko, the outspoken peasant who helped many Jews and definitely saved five, says: *I do what I think is right, not what others think is right.* Arnacka, the simple pious woman who, during the war, selflessly helped many Jews, expresses similar sentiments: *I am not interested in the opinion of other people. I am not concerned with what others do. I was always independent. I am not interested in my neighbors, never bother with gossip. I only attend to my own business. I have to be busy and do not like to have too many people around me. What others do does not interest me and I am never bothered by what they think about me.*

It seems that Jan Gonski shared with other rescuers only their protection of Jews. He was a known painter, a distinguished professor of art, and a member of the academy of arts. During the war he helped a Jewish friend and worked in the Council for Aid to Jews. How does he see himself as far as independence and moral values are concerned? *Whatever I do I am compulsive about it. It has to be done properly. Besides I have the courage of my own convictions. I have to be able to face myself. Had I not helped, I would have found it unpleasant to be alive. I had to comply with my own rules, to my own expectations, and they ordered me to act. As for other people, I do not bother with what they think about me. When I undertake something I stick with it. When I do something it has to be done properly. I am a perfectionist. With perfectionists it is an unending process. Once I decide that something is right, this is it. I have few contacts with people in general. I have no time. Every person depends on the opinion of others, but compared to other people I depend less than they. Of course I like when people approve of me but, when they do, I am only pleasantly surprised.* Gonski's answers show how much in common he had with the other rescuers.

None of the rescuers I interviewed directly had trouble talking about self-reliance and the need to follow personal inclinations and values. This need to be independent of others also appears as a strong characteristic among those in the large sample. Nearly all rescuers saw themselves as independent[32] and as unmoved by others.[33]

The Jewish survivors I questioned personally all characterized their rescuers as independent and as being motivated by personal values. The quality often mentioned in the testimonies and memoirs of survivors, and one that comes close to independence, is the rescuers' courage. The overwhelming majority (85 percent) of the Jews described their helpers as courageous. In sharp contrast to such descriptions about rescuers are those of paid helpers. Very few (12 percent) perceived their paid helpers as courageous.

Among the survivors I interviewed, only one of the paid helpers mentioned was described as independent: the peasant who saved Lola Freud. However, no survivor characterized paid helpers as being propelled by personal values when they protected Jews; all were described as having been motivated by money only. At one point I asked Lola Freud: "Do you think this peasant would have continued to help you had you run out of money?" Her immediate reaction was: *I don't have to think, I know he would not have kept me . . . he would not have kept me because he was only money motivated . . . he kept raising the price all the time.*

To sum up, despite the rescuers' heterogeneity and wide range of social backgrounds, they show some consistent characteristics and motivations. Central among these is their inability to blend into their environment, a condition I refer to as individuality.

Individuality was closely related to rescuers' independence, which was also related to their ability to follow their own moral imperatives, even when these were in opposition to the values and norms pervasive in their environment. The close connection among a rescuer's individuality, independence, and the effectiveness of personal morality was evident in most cases.

While only a few of these Poles were aware of their individuality practically all of them were aware of their independence. Connected to this was also their awareness that they were propelled by moral standards and values not necessarily shared by others, standards that did not depend on the support and approval of others but rather on self-approval. Again and again they repeated that they had to be at peace with themselves and with their own idea of what was right or wrong.

Closely connected to the rescuers' moral convictions and values was their long lasting commitment to stand up for the needy. This commitment was expressed in a wide range of charitable acts that extended over a long time.

What was the basis of this pattern of behavior? How was it perceived by the Jews and Poles? Invariably the survivors presented a glowing picture of their protectors by almost unanimously defining them as "good natured" and by describing their efforts on behalf of the needy as limitless and enduring. Thus in a highly emotional voice Pola Stein said that: *Before, during, and after the war anyone who approached him was never denied help. Anyone could eat and sleep in his house. It was not only Jews, it was anyone, any tramp, any dog, whoever came to him, he felt compelled to help. Throughout his life he was caring for the poor, the vagabonds, the sick, and all his neighbors kept asking him for favors. People often took advantage of him but he never changed. I never met anyone like him again.*

Similarly, Danuta Brill saw her helpers as *. . . very, very noble. They had incredible insight and sensitivity, sensitivity to the suffering of others. . . . Their help was extended to everybody who needed it, not necessarily Jews.*

Nor were the few outspoken anti-Semitic rescuers evaluated differently. Indeed, David Rodman called Hela Horska his second mother, explaining that by saving him she had given him life for a second time. Like most survivors, he emphasized his helper's sensitivity to the suffering of others. He described his helpers as extremely tolerant of those who turned to them for aid. They were interested in helping, and that was all that mattered: *They would help anybody. They were both active in the underground, trying to do good for everybody. They were sometimes trying to save Poles from the clutches of the Gestapo. Sometimes they were successful. Once, a Jewish woman, a stranger, came to the house and said that unless they kept her overnight she would give herself up to the Gestapo. She insisted that she could not take the torture and the suffering she had been through, that she had nowhere to go. Now she claimed she had to have a place to rest, eat, and sleep. She believed that she would later find other Poles who would take her in but at that point she had to stop over in this place. She could not even think straight. They took her in.*

Poles describe their help differently from their charges. Whereas the Jews were glad, even eager, to praise their protectors, the rescuers were reluctant to talk about their aid. Even those who did, spoke only in timid and restrained ways. As a rule, I had to prod and probe before any of them mentioned those aspects of help that put them in a particularly favorable light. Instead, they consistently underplayed the risks and sacrifices inherent in this kind of aid. These rescuers succeeded in projecting a very modest image.

Efforts to underplay and deny the importance of rescue were closely intermingled with all other mentions of helping behavior. Often it was hard to sort out the various descriptions. Nor did these Poles admit easily to the presence of a persistent pattern of charitable actions.

To illustrate, when Ada Celka spoke about her help she carefully avoided using superlatives. Listening to her I knew that she was holding back. Only after considerable coaxing did she lose some of her reserve and admit that besides saving one Jewish woman she also sheltered Russian partisans in her one-room apartment. Toward the end of our meeting she explained: *By saving the Jewish girl I simply did my duty. What I did was everybody's duty. Saving the one whose life is in jeopardy is a simple human duty. One has to help another regardless of who this human being is as long as he is in need, that is all that counts.*

Another rescuer, Maria Baluszko, at first also resisted telling me that her aid to Jews was an extension of a tradition that involved helping the poor and the destitute. When I touched upon her reasons for rescue she was at a loss. Then, instead of answering, she asked: *What would you do in my place if someone comes at night and asks for help? What would you have done in my place? One has to be an animal without a conscience not to help.* I had no answer. Impassively I waited for her to continue. Only at that point did she tell me: *In our area there were many large families with small farms; they were very poor. I used to help them; they called me mother. . . . I used to help all. . . . When I was leaving the place people cried. I helped all the poor all that needed help.*

Like these others, Hela Horska preferred not to dwell on her help. Finally in response to my prodding she said: *All my life I worked for social causes—I was moved by God's will. It did not matter who it was if someone needed help I had to give it. . . . I helped because a human being ought to help another. I also did it for religious reasons. . . . I believed that the world became such a horrible place and someone had to counteract the cruelties of this world. . . . I always helped others. During the war I helped many Jews.*

Similarly, after close questioning I learned from Adela Uszycka that she had a long history of good deeds and that her aim in life was to help and save others. She worked in different capacities and it mattered little who the needy were. Already during the 1939 bombings she took care of the wounded. After the fall of Poland, she joined the Polish underground and as a courier was sending Polish soldiers over the border to Rumania. Later on, as a member of the Council for Aid to Jews, she continuously worked on behalf of Jews. At the time of our meeting in Warsaw, she was 87, and barely able to move, even with the use of a walker. As I was about to leave she hugged me and said: *Oh I want to leave this body. I want to go there [pointing to heaven] and there I will also do good for people. With this sick body I can do nothing here.*

Deeply religious, Janina Morawska at first credited God for all her good deeds. When I insisted on a more personal account, she referred to herself as an altruist explaining: *I feel for other people and always did. At the beginning of the war as our poor soldiers were running away from the enemy, my mother and I would distribute food to them. . . . Later on I helped Poles who worked in the underground. I warned them about imminent danger. . . . I was also sending packages to prisoners of war. These were Poles who had no families that could look after them.*

Additional quantitative support for the long-lasting commitment to help comes from the entire sample of helpers. Sixty percent of them had a past history of charitable deeds. For the remaining 40 percent, this kind of information is unavailable.

What do such repetitive and enduring actions mean? Why do the rescuers deny the exceptional quality of their life-threatening protection of Jews?

As a rule we take most repetitive ideas and actions for granted. What we take for granted, we usually accept. Furthermore, what we take for granted and accept, we rarely analyze or wonder about. In fact the more firmly established patterns are, the less likely we are to analyze them and think about them. In a very real sense, therefore, the constant presence of, or familiarity with, ideas and actions does not mean that we know or understand them. On the contrary, the customary patterns are accepted and taken for granted. This in itself may impede rather than promote understanding. Closely related to this tendency is another one: that which we are accustomed to and tend to repeat doing, we have a hard time seeing as exemplary no matter how otherwise exceptional it may be. In a real sense, therefore, immunity to seeing the special and extraordinary leads to missed opportunities.

In what way and to what extent do these broad assumptions apply to the willingness to risk one's life for Jews?

Findings thus far presented point to a continuity between the rescuers' past history of charitable actions and their protection of Jews. That is, risking their lives for Jews fits into a system of values and behaviors that included helping the weak and the dependent. This analogy, however, has serious limitations. Most disinterested actions on behalf of others may involve inconvenience, even extreme inconvenience, but only rarely do they even approach the possibility that the giver may have to make the ultimate sacrifice—his or her own life. For these particular Poles, then, only during the war was there a convergence between historical events demanding ultimate selflessness and the already established predisposition to help.

What special effects, then, did this past history of aiding the poor have on the rescuing of Jews? I have already hinted that there seems to be a relationship between the tradition of helping and the protection of Jews; this in turn was closely related to the rescuers' modest appraisal of these life-threatening acts. This modesty on the rescuers' part may be traced to their acceptance of helping the weak and needy.

Though expressed in a variety of concrete ways the matter-of-fact attitudes toward rescue were consistent. Thus, Janina Mozawska explained: *Among the Jews who were visiting me at night was a two-and-a-half-year-old child. They lived in the forest near to my house. I would let them in at night and give them food. When it was cold the child had frostbites, wounds. . . . I could not watch it. In cold weather I would let them stay with the child for a week at a time; they would not go out. But it was dangerous. I lived in one room. . . . Still, at that time I could not think about personal well-being. . . . It never occurred to me that one could behave in a different way. I just had to help people who needed help and that was that. I was always ready to help the needy, always. That is how I am.*

Jewish survivors were very aware of their rescuers' feelings of duty. One of them, Abram Levi, met his protector only after she agreed to shelter him. This is how he saw her: *She was a very decent, honest woman. She did not think much about what she was doing. She helped because as far as she was concerned there was no other way. While I stayed with her, she also took in my brother-in-law with his wife when their apartment was endangered. At one point her neighbors complained that she kept Jews and that she was endangering all of them. In fact, they threatened her. Eventually the Gestapo came to her apartment, but they did not find us because we had a good hiding place. Even after the search she wanted us to stay, not because she did not realize the danger, but because she knew that we had nowhere to go. But we were scared to remain there. At a friend's suggestion we went to the Hotel Polski, supposedly for a few days until we could find a suitable place to stay. She then came to the hotel begging us and imploring us that we should move back to her place, that the hotel was unsafe. She was right, but we did not listen. From the hotel the Nazis indeed sent us to Auschwitz.*

This matter-of-fact attitude to Jewish rescue also became apparent from information offered by the entire group of helpers. Most of them (66 percent) perceived their protection of Jews as a natural reaction to human suffering, while others (31 percent) insisted that to save lives was nothing exceptional. In contrast, only 3 percent would characterize the saving of Jews as extraordinary behavior.

An overall refusal to perceive the drama of these life-threatening and risky actions was expressed in still other ways. Some Poles omitted from their accounts events that attested to particularly noble and courageous aspects of their actions. This tendency is apparent from a comparison of information collected from pairs of rescuers and rescued.

Ada Celka again provides a vivid illustration of this point. To my suggestion that keeping the Jewish girl Danuta must have entailed economic hardships, Ada reacted with a flat denial. She also failed to tell me about a few facts that would have enhanced her image. I heard only from Danuta that Ada had planned and almost succeeded in smuggling Danuta's parents out of a working camp and placing them with a Polish family in the country. This, according to the daughter, involved extraordinary efforts. Ada was not an influential person; she had few connections and no money. Her success in locating a peasant family willing to protect Jews could be ascribed to her willingness to try again and again and to her strong determination. Finally, all was ready, and detailed plans for smuggling the parents out of their working camp were set in motion. On the chosen day Celka went to the appointed place next to the camp, but she waited in vain. The day before Danuta's parents had been deported to a death camp.

Ada also never bothered to tell me that when food was scarce, which it often was, she fed her invalid father first, and then Danuta. She and her sister ate only after her father and the girl had had enough.

Another example of a significant omission comes from a simple Polish woman who found a Jewish girl and, despite all threats and dangers, cared for her lovingly. At one point the woman was denounced for harboring the little fugitive. When the police came to take the child away, the distraught protector pleaded with them to shoot her instead, and let the child live. The men took pity on her and left, leaving the girl behind. This incident was missing from the woman's account and was reported only by outside witnesses.[34]

More frequent and more direct expressions of this need to minimize the exceptional qualities of rescue relate to comments about heroism. Indeed, not only did most helpers deny that aid to Jews was heroic; they became embarrassed when this possibility was suggested to them. This was the case with Staszek Jackowski, a recipient of a Yad Vashem medal, who saved thirty-two Jews by hiding them in a special place built under the floor of his living room. He did this without the knowledge of his mother and sister, who shared his home. Jackowski felt uneasy when the interviewer pointed to the heroism of his actions and insisted: *I don't consider myself a hero. Just a man who did his simple duty in opening his home to Jews.*[35]

Stach Kaminski, the prominent underground figure, said that for him Jewish rescue . . . *was not an emotional call. I had no desire to sacrifice my life or perish. . . . I was lucky that the Germans never tortured me. Heroism starts when there is suffering.* Kaminski argued that the mere possibility of suffering does not imply heroism.

Upon our meeting and before I could ask her anything, Eva Anielska said: *Now those who investigate the problems of saving Jews want to make heroes out of us. But we are not heroes, no great humanists. Nor were we prompted by great ideologies.* Then she proceeded to say that her help started gradually, only after she had gone to the ghetto to visit her dear friend. When she saw what was happening in the ghetto, she widened her efforts. Eva emphasized that, without the cooperation of others, she could never have done as much as she did. For her the unpremeditated and gradual nature of rescue, together with the fact that others cooperated with her, was reason enough to remove it from the realm of heroism.

To this day Paweł Remba limps from an injury that occurred when he smuggled Jews out of the Warsaw ghetto during the uprising. For this and other acts on behalf of Jews, he was awarded the Yad Vashem medal. When Paweł and I met in Israel he categorically denied that he or others like him were heroes: *I would absolutely not make heroes out of the Poles who helped. All of us looked at this help as a natural thing. None of us were heroes; at times we were afraid, but none of us could act differently.*

Like Paweł most rescuers tried to dissociate their actions from heroism. Many did it by pointing to their fears and anxieties.[36] The underlying assumption in such arguments was that fears were incompatible with heroism.

Felicja Zapolska was one of those who emphasized fear. She felt that: *In general, those that helped were sensitive people who tried to overcome their fears. Everyone was afraid. Do not believe it if someone tells you that they were not afraid because it has to be a lie.*

Stach Kaminski, too, had no trouble pointing to his fears. He was convinced that this was a universal sentiment which others shared with him: *I was not more afraid than most. Every situation that is not anticipated causes some fear. I had horrible dreams that I would not accomplish what I was supposed to. . . . I feared that, and I feared the Nazis.*

Even though Stefa Krakowska saved fourteen Jews and helped many others, she referred to herself as a coward. She backed up this view by pointing to her fears. When I suggested that hiding Jews required courage she asked: *How could I be a coward and a hero?*

Because Maria Baluszko's house was exposed, she sent five Jewish fugitives to her relative who lived alone close to the forest. She took over the responsibility of feeding them. Baluszko explained: *Since he knew the Jews, he did not mind keeping them in his house. But soon there was a raid during which he [the relative] was caught and sent off to Germany for work. The Jews remained in the hiding place in his pigstall. It was a cellar built especially for them. One of them slipped out and came to our place to tell us what had happened. . . . I had servants, little children, and an exposed house. It was dangerous. So I sent two of them to someone I could trust. For three I made a hiding place in my own pigsty. Then German soldiers, together with Lithuanians and Latvians, decided to stay in our place. This was dangerous. They [the Jews] could not continue staying with us. In the nearby forest, there were some bunkers left from World War I. We took them to the bunkers, and supplied them with food. . . . I was fearful all the time, but especially when the German soldiers came to stay with us. I became tense and nervous. I was very much afraid. As far as I am concerned, one has to help all that need it. But I was nervous. In fact, my nerves were shattered; I would scream and yell for no reason at all. I had terrible outbursts, during the war and right after.*

This feeling of nervousness, however, was not universal. Some pushed their fears and anxieties into the background or managed somehow not to show them. The writer Roman Sadowski remained unaffected by his fears: *Of course I was afraid, but not too much. I was so overworked that I had no time to think about it. . . . I was so absorbed in my work that I did not think.*

Similarly, of the entire group of Polish helpers, half mentioned a variety of fears, all related to their protection of Jews.[37] In contrast, as noted earlier, the survivors refer to only a few of their helpers as fearful.

Fears were usually discussed together with dangers. The rescuers mentioned either one or both when they wanted to remove from the description of their protection of Jews the aura of the unusual and the dramatic. At times they insisted that more important considerations pushed both fear and danger into the background. What were they? Eva Anielska explained: *I was afraid, but in moments of danger one did not think about it. It was so important to give the help that one tried not to think about the danger. . . . Now, from a distance, it looks too dangerous, but then stronger than fear was the desire to help.*

An important technique of minimizing the dangers was to place them within the context of other threats. That is, by pointing to the continuous existence of other dangers, these Poles tried to make the protection of Jews appear less perilous. In a variety of ways, then, they tried to underplay the risks that were an inevitable part of Jewish rescue. What were some of these ways?

Roman Sadowski argued that his involvement with Jews might have helped him avert disaster. This he felt happened because as a rescuer he became particularly alert. He explained: *Danger avoided me in an extraordinary, miraculous way. When Germans were catching people, I usually appeared some minutes after it was over. Or I came too early. . . . Besides, the Germans were persecuting all of us. One never knew, one was never sure what would happen next. Even people who had nothing to do with illegal activities could be caught in an everyday roundup and sent to Auschwitz, and there they would simply die. For no reason at all one could perish, and there were many people like that. In fact, maybe those who were engaged in some kind of anti-Nazi activity were less likely to be caught because they were more cautious, more aware. We were prepared and trained.*

A different perspective was offered by Stefa Krakowska who insisted that *one cannot make a big deal out of it [rescue]. It was wartime and one was risking. You think that I was not afraid? All the time I was! . . . But I felt threatened with and without the Jews. I could have been caught by the Germans during a raid for no reason at all. This was fate. If I succeeded, all was well; if not, what could one do? It was not heroism at all. I was scared, very much so . . . sometimes I was exhausted emotionally. It was too much for me. Till a few years ago I had nightmares; I cried at night.*

Tomasz Jurski also emphasized that help to Jews represented only an addition to the already existent danger. He felt that aiding Jews was less risky than it appeared: *None of us analyzed it coolly or rationally. Maybe if I were asked to do today one-tenth of what I did then, maybe I would be paralyzed by fear. In those days there were different conditions. One looked differently at life. Life was worthless. One could simply pass in the street and for taking out one's hand from the pocket too slowly one could be killed. I saw something like this happen; a man was shot because to an order to take his hands out of his pockets he acted too late. We lived on borrowed time. There was no guarantee whatsoever that one would survive, regardless of whether one followed the German directives or not.*

Stefa Dworek, although much older and from a more humble background, echoed Jurski's assertions: *One could go out on the street and never come back, without doing anything. Many times I had to run away because they were catching people. This was happening all the time, one was never safe. Anyway this was war, one never knew what was awaiting one.*

Stach Kaminski argued that Nazi threats to the entire population facilitated aid to Jews. In creating threatening and dangerous conditions, in spreading terror among the entire population, the Nazis were in effect making it easier, at least for some Poles, to oppose them: *After all, if one could be punished for anything at all, or nothing, then one might as well do something worthwhile. Moreover, the oppression itself gave rise to a spirit of opposition, an urge to oppose those who were inflicting such suffering.*

Some rescuers, without denying the ever-present perils, consoled themselves that at least if they died for saving lives it would have been for a worthwhile cause.

Paweł Remba reasoned: *I was often afraid. Courage is a relative thing. One always hoped that there would be no tragedy. I felt strongly that this was what I had to do. . . . One could perish anyway. One could die for anything. The Jews were persecuted: to die for helping them was not bad. It would have been an honor to do so. If I were caught for smuggling and*

*shot, this would have been a dishonorable death. But to be shot for saving
a human life is a different matter.*

In a variety of ways, then, the rescuers tried to remove their aid from the
exceptional and extraordinary. Some did it by describing it as a mere duty.
Some pushed the dangers into the background; some emphasized the great
value of saving a life; some tried to diminish the danger of their actions by
seeing it as simply another part of a threatening environment.

How realistic was the insistence that the danger inherent in the protection of Jews was only a part of other equally perilous conditions? What
kinds of adverse experiences did these Poles in fact have both related and
unrelated to Jewish rescue?

It is true that some of these helpers lived through tragedies unrelated to
the protection of Jews. Such experiences only attest to the shaky and
threatening conditions of the times. But were the sufferings and punishments that could be inflicted on these Poles for Jewish protection the same
as those unrelated to this help? Were the helpers right that saving Jews
was just one other danger among many?

Jurski's experience seems to confirm this possibility. He was caught in a
routine roundup, unrelated to any illegal activities: *One day as I was
passing through the streets of Warsaw I noticed that there was a raid; the
Nazis were catching passersby. I had a lot of money on me because I was
going to buy weapons for the underground. I began to run. They caught
me and took the money away. Then they brought me to Sucha (Gestapo
headquarters). They interrogated me about political things, not about
Jews. I realized that I might break down. I was afraid. I knew that my
family was home with fourteen Jews. It was tragic. I was sure that if I
started talking then many people would die. I did not know what I would
do if I broke down. Thank God I did not break down. The Nazis tortured
me but I managed to keep quiet.* Jurski was sent to Auschwitz, where he
stayed until the end of the war, for almost two years.

Zofia Staniewska and her family sheltered a young girl whose Semitic
features spelled disaster. When the child was brought to them by the
distraught mother, a mere acquaintance, they could not refuse her and
decided to keep her in the back of the apartment, in the maid's room,
where they felt she would be protected from inquisitive eyes.

The girl proved intelligent and cooperative. All day long she sat in the
back room reading and memorizing Catholic prayers. She never complained and was ready to do anything she was asked to. Staniewska remembers: *We felt sorry for her. She looked so scared and tried so hard to
please that she made us feel guilty by her very agreeable and compliant
behavior. . . . My entire family was deeply engaged in underground activities. I had a teenage brother who was active in sabotage work. But even
though we were very busy during the day, we tried to find some time
during the evening for our little guest. I used to play with her, but she
never really laughed like a child. She was always on guard not to offend,
not to impose. Sometimes in the evenings I took her for walks. This I could*

do only when it was dark. Even then I would put up the collar of her coat, and try to cover as much of her face as possible. If anyone had set eyes on her there would have been no doubt that she was Jewish.

The child stayed with them for a few months. Then they realized that the Polish police were spying on them. They became aware that their apartment was being watched. One day Staniewska's mother was stopped on the street and her documents were checked: *We had to talk the situation over; we all knew that something was brewing. My brother, who had a golden heart, became convinced that it was dangerous to keep this child and that someone must have seen her. My father was never in favor of keeping her, but he went along with us. Now, with the danger so close by, he insisted that the risk was too great. He was a realist; he saw things more clearly and perhaps this is why he was more afraid. I wanted to keep her. I felt terrible pity and compassion for her, but my voice did not carry. When the mother came to visit, we told her that she must take her daughter. "Where will I go? Where can I take her," she asked, with disappointment and pain. We explained that it was too dangerous to keep her. She became angry and said, "You think that it is only us; there will come a time for you too."*

Miraculously both mother and daughter survived. The child became a known journalist and keeps in touch with Staniewska. The two, in fact, became great friends and the former little girl remembers her benefactors warmly and is grateful for the few peaceful months.

Ironically, however, danger did not elude the Staniewski family. It came for no reason at all with disastrous consequences. Unrelated to the protection of Jews or underground involvement, the young son was caught in a routine roundup. He was taken to Pawiak prison, and without the Nazis ever discovering whether he had "sinned" against them or not, he was executed with others who died also for no apparent reason.

A similar tragedy befell Piotr Wrona, an engineer I interviewed in the Jewish Historical Institute in Warsaw. As one of the few Poles who worked in the ghetto, he was able to smuggle food. For some Jews Piotr made false documents and to some he had offered shelter in his apartment. The longest, however, any of the fugitives had stayed with him was twenty days. Wrona explained: *It was dangerous in my apartment; it was exposed. I also had inquisitive neighbors. . . . Nor could I belong to the underground because I had a family whom I had to protect.* Still, in the end, he did not succeed in protecting his family. His son, his only son, a gifted youth of 17 was murdered by the Nazis. How did it happen? It was on one of those routine roundups in which innocent people were caught and shot. Piotr lost his son, not because he was helping Jews, not because he belonged to the underground, but only because his son happened to be in the wrong place at the wrong time.

In a sense, then, the rescuers were right. Chance played its tricks and people did perish for no transgressions at all. Disaster was likely to come from any direction and without provocation. Still, to engage in actions clearly defined as "crimes" certainly increased the possibility of disaster above and beyond mere chance. Whereas some Poles had died for no "crimes" at all, those who were saving Jews were knowingly courting disaster. Because of it, they were more likely to have suffered. When the entire group of rescuers is examined, it is clear that most of them were at one time or another threatened with a variety of disasters including arrests, denouncements, and blackmail: a minority of helpers were actually arrested and imprisoned because of Jewish rescue, and only very few succeeded in escaping from such experiences.[38]

In contrast, adverse wartime experiences unrelated to the protection of Jews were less pervasive. Less than half of the rescuers were exposed to dangers unrelated to Jews.[39] These were the realities that some Christians chose to disregard. They knew about the perils, but in their minds succeeded in minimizing them by removing from Jewish rescue the aura of the exceptional and the extraordinary.

Closely associated to the long history of charitable acts and the refusal to see their aid to Jews as extraordinary was their inability to explain their aid. This was true even for the more educated rescuers whom one would expect to be more introspective. Moreover, taken for granted, perceived as a duty, not easily explained, the rescuing of Jews had usually unplanned beginnings. Hard to disentangle from the matter-of-fact attitudes and the inability to explain, this unpremeditated aid could happen gradually or suddenly.

Indeed, even the sophisticated and prominent painter, Jan Gonski, was at a loss when asked why he risked his life for Jews. At first he refused to see his actions as special. When pressed, he answered by telling a story. During World War I as a soldier at the front he was caught in a heavy battle. It was a gloomy dark night, and the bullets were flying above his head and all around him. Clinging instinctively to the soil underneath his stomach, he soon realized that the only way he could possibly come out of this alive was to reach a nearby ditch. As he was about to crawl to safety, he heard faint whisperings that came from an injured comrade. Fatally wounded, unable to move, too weak to communicate in any other way but whispers, the man pleaded with Gonski not to leave him alone. He was afraid to die. As he listened, the thought crossed Gonski's mind that he could be of no use to this unfortunate man and that staying with him would only endanger his own life; yet he could not leave him. He simply could not refuse the comfort his presence might have given to the dying soldier. For him there was no other way. Then a bullet hit Gonski, inflicting a severe wound.

The rescuer Marek Dunski also tried to explain his actions with a story. During the Warsaw uprising as a prisoner of war he and a friend were asked to carry a wounded Russian soldier. The soldier was heavy; they were hungry and exhausted; and therefore they kept tumbling and stopping for rest. Seeing their difficulties and discomfort, a German asked them to unload the wounded man and place him at the side of the road. As they did, almost in a flash, it dawned on them that the German intended to shoot the Russian. This is how Dunski explained what happened: *I am usually a quiet person. Nor do I have any special fondness for Russians. But at that moment, both my friend and I went almost crazy. We started to scream and object so violently that the surprised German called a car into which we placed the wounded soldier. At that point we just reacted, automatically; both of us must have felt that a human being ought to be saved at any price. So you see the situation itself brings out certain strong feelings and reactions. Who knows, who is to tell? In the same way I had to save the Jewish children that I did, I just had to.*

Gonski and others emphasized the inevitability of their reactions, which they traced to their need to stand up for the weak and helpless. In part this very acceptance of their inclination to protect the needy interfered with their ability to explain their aid.[40] It was something they had to do, leaving no room for elaborate discussions of motivations, only indirect indications of what these could have been.[41]

Given the matter-of-fact acceptance of rescue, it is not surprising that this aid often began in a spontaneous, unpremeditated way. Unplanned and unanticipated, this participation could start either gradually or suddenly.

Gonski's example illustrates the unplanned and sudden reaction to the helplessness of others. Another dramatic instance of the same behavior was offered by Janka Polanska, the young woman who saved ten Jews. Janka could not resist the temptation to help the needy. This is what happened: *Once I was passing through the streets in Warsaw when I saw a man who reminded me of my father. He was disheveled, distraught, and seemed lost. He looked and acted strangely. He simply stood in the middle of a busy street crying piteously. As I came closer, I recognized him as a known Jewish judge. I took him by the hand and said: "Father, come with me, do not cry." He had just witnessed the arrest of his mother by the Gestapo. Instead of helping her, he had run away. He was totally broken and felt terrible remorse, even though obviously he himself barely escaped being caught. As if in a trance, he let me lead him away. I took him home, put him into the hiding place with the others. He refused to eat. This was the first time I had to insult a person like this. I swore at him and cursed him. I was afraid that, should he die, then we would all be in great trouble. What would we do with the body? Everybody was very nervous, irritable. They were stuck in this little camouflaged room; they were all touchy. Naturally, they were all angry at me. I brought him there. He stayed for a few weeks; then I found him another place.*

For over two years Jan Rybak sheltered Pola Stein and her father simply because they came to him and asked for help. He stood out within his environment. He was the shortest man in the village yet his good heart was enormous. He was unusual in other ways, too. While his formal schooling did not extend beyond three years, his informal education was extensive; he read whatever he could put his hands on. His knowledge and love for learning gained him the nickname "philosopher." Three months before the end of the war he came back from a trip to town with a Jewish girl whom he saved from execution. The girl stayed with them till the end of the war. How did it happen? As he was passing the marketplace he heard screams, "Jude Jude." He soon noticed two Germans dragging this Jewish girl. He recognized her as the daughter of a local storekeeper. When he realized that she was about to be shot, he went over, grabbed her by the hand, and said: "This is not a Jude, this is my cousin," and left. All those around him were too stunned to react. Asked why he did it, he could not explain.

This unplanned and sometimes sudden way of extending help only underscores the rescuers' need to stand up for the poor and helpless. So strong was this need to help, so much was it a part of their makeup, that it overshadowed all other considerations. It seems, indeed, that it mattered little who the needy were. In a sense, the help was not given to a particular person, but to the one who was believed to be in greatest need.

Asked why they had saved Jews, the Poles emphasized over and over again that they had responded to the persecution and suffering of victims. What compelled them to act was the persecution, the unjust treatment, and hence the people's need for help and protection.

This ability to disregard all attributes of the needy, except their helplessness and dependence, I refer to as universalistic perceptions. Evidence for the presence of such universalistic perceptions of the needy comes from a variety of sources. Turning first to the quantitative evidence, practically all helpers believed that it was the terrible plight and helplessness of the persecuted that mattered and not their Jewishness.[42] The survivors reconfirmed this assertion.[43]

What did such universalistic views of the victims entail? How were they explained?

The rescuer Leon Wronski told me that in 1940, before the Final Solution was put into effect, the Nazis had also persecuted the Polish intelligentsia and at that time he had focused on helping Poles. Only when the Nazis began to concentrate on Jews did Leon's help also change direction! *In 1943 it was clear that the worst off were the Jews. So one had to give help where people were most helpless; this was clear.*

Even though Stefa Dworek was very different from Leon, her reasoning was similar to his. Asked why she continued to keep in her one-room apartment an uninvited guest, a Jewish woman, over the objections of her husband, who eventually left her, she answered: *I knew I could not let her go. What could I do? Even a dog you get used to and especially to a fine person like she was. I could not act any other way. . . . I would have helped anyone. It did not matter who she was. After all I did not know her at first, but I helped and could not send her away. I always try to help as best as I can.*

Bolesaw Twardy was and remained a highly principled idealist. Reflecting on his continuous and selfless help to Jews, Twardy emphasized the abstract principle that prompted him into such actions: *I looked at the war as a time that one had to fight with the anti-humanitarian phenomenon. Emotionally I was deeply involved and opposed to the ideology of Hitlerism. . . . When I thought that someone because of his nationality or origin could be killed, I became infuriated. Because of this I had to help. After all, the Jews were the most helpless people; they had no chance to fight, even before they could fight, they were condemned to death.*

Aron Blum, too, as he described the courageous deeds of his benefactor Franek Dworski, insisted that the suffering of the Jews prompted him to help. Still Blum believes that to respond the way he did Dworski had to be a very special man. *He is an unusual man. He is unusually sensitive and good and eager to help everybody not only Jews. He was willing to help anyone, no matter who they were. He looked upon the help as his duty as a natural thing. He simply could not stand the suffering of other people.*

Wojtek Kominek, the self-possessed underground figure, known for his many courageous escapades also believed that when helping Jews he was moved by their plight: *Seeing the horrible misery of people I felt compelled to help. I did not much think about it; I did it as a matter of fact. One had to do it, so I did it. Those who were in greatest need had to be helped.*

Dr. Estowski was deeply involved in helping others, Jews and non-Jews. He helped both as a member of the underground and as a private citizen. How does he describe his help to Jews? *Whoever came to us we always managed to help. I felt that it was my duty to help people. It was not because they were Jews. I had a simple obligation to help people. One had to help people. It was not for us a question of them being Jews or not, just anyone who needed help had to get it. Jews were in a specially dangerous situation; all of us who were helping were aware of this fact—that because of their difficult situation, they had to be helped the most. After all, a Pole could somehow help himself, but the Jew was in a more horrible situation and could in no way help himself.*

Asked why her protector, Roman Sadowski, concentrated on saving Jews, Sara Federer answered without hesitation: *He felt most sorry for Jews because they were more persecuted than anyone else; he knew what the situation was. . . . He was very upset and suffered for the Jews. He was too humane, so warm, so considerate. He is a rare person. . . . So much love for people one rarely sees.*

In less flowery terms, Sadowski reconfirms what Sara had said. *Their being Jewish did not play a part at all. Regardless of who they were, needing help was the criteria. . . . Sure enough, during the war I focused on helping Jews. Each of us knew that they were in the worst situation. It did not matter that they were Jews. We looked upon this help as our duty, as an obligation. All else did not matter. We had to do it. Human life was at stake.*

Anielska, too, when commenting on her own help to Jews and that of others, stresses how unimportant the Jewishness of the victims was: *One saw the Jew, not as a Jew, but as a persecuted human being, desperately struggling for life and in need of help . . . one did not have a Jew before one's eyes, at least not the Jew that one was accustomed to think about. Instead, it was a persecuted, humiliated human being asking for help.*

It mattered little that they were total strangers. For example, Aron Blum came to Franek Dworski as a stranger, a haunted man, without papers, without money, without prospects, and without hopes. As soon as he knocked at the door, he was welcomed warmly. Blum recalled: *When I came to his house he did not know me at all. In fact, I did not even know the man who directed me to him. As a welcome he said, "Bread in this house you will not miss." He was as poor as I was, but he shared all. He simply looked upon this help as the most natural thing. He had so much heart and courage. He made papers for me, found me a job, and gave me shelter. After a while, and upon my request, he traveled to Lwów to bring a friend of mine, also a Jew. To make this possible for my friend he arranged false papers. On the way the train was searched by the Nazis. It was customary in such situations to throw away all incriminating evidence. Not he. He hid the papers. Dworski was not afraid, even though he knew that he could die for it. But he also knew that, without these papers, he could not bring my friend back, and that this in turn could cost my friend his life. Dworski had courage and luck. All went well. He brought my friend, whom he kept in his house as he did everyone else who turned to him. All this he did without ever receiving, or expecting to receive, anything in return.*

Indeed, as I have shown elsewhere, only a minority of rescuers limited their aid to Jewish friends, while the overwhelming majority risked their lives either for total strangers or mere acquaintances. This underscores the fact that to these Poles it mattered little who the persecuted were. Anyone in need qualified for help.[44]

The compelling moral force behind the rescuing of Jews, as well as the universal insistence that what mattered were the victims' position of dependence and unjust persecution, combined to make such actions universalistic. In a sense, it was this moral force that motivated the rescuers independent of personal likes and dislikes. Some of them indeed were aware that to help the needy in general, and the Jews in particular, one did not necessarily have to like them. Liking and helping, they knew, did not necessarily go hand in hand.

Gentle Celka who expressed a deep compassion for the suffering of others emphasized the difference between help and personal attraction: *I would help anyone, anyone who needs help, but this does not mean that I like everybody.*

Others moved a step further by insisting that if those in need were people they disliked they would still protect them.*

The prominent economist Zygmunt Rostal was aware of the impersonal

---

*This resembles closely the pattern shown among the anti-Semitic helpers, who never modified their opinions about Jews.

quality of rescue: *It was something so natural for me that I simply did not think about it. I did not dwell on it. What was important was that one had to help, it was natural . . . now too I would not think about it. When I see persecuted people that are helpless, I defend them regardless of the consequences. This is how I am. I do not dwell on it. If this help was needed by a French, a Spaniard, and even a German, I would also give it. For me it was a human being that was in need of help. A human being that was totally helpless.*

For Rostal to include a German among those who would receive his aid meant that he could not turn away even those he disliked. The same sentiment was more explicitly stated by a director of archives who, with great difficulty and under considerable danger, built a hiding place for his Jewish employee and her family. Like most of these Poles he, too, was a sensitive man who identified strongly with the suffering of others. But it was a sense of duty that compelled him to act rather than a personal attachment or warm feelings. Fully aware, he explained: *Even if you were the worst kind of people, I would have considered it my duty to have saved you.*[45]

The idea of helping those one disliked went beyond the hypothetical. Jan Rybak readily admitted to his dislike for the Russians, yet to escaped Russian prisoners of war who turned to him for help he never refused food and shelter. He could not act any other way. He had to help everyone regardless of who they were.

Marek Dunski, too, after he told me how he refused to obey an order to put a wounded Russian soldier into the ditch so that he be shot, almost in the same breath, assured me that he disliked Russians.

Rybak's and Dunski's dislike extends to a category of people: the Russians. Such a dislike on the part of helpers could apply personally to those for whom they were risking their lives. Leon Wronski was aware of this fact: *With some of the Jews I was helping I became friendly, with some I did not, others I felt repulsed by. But this did not affect my action. I was still helping. I helped all—those I liked and those I disliked.*

Felicia Zapolska helped Jews both as a member of the underground and as a private individual. For two and a half years she kept in her apartment a Jewish woman, a stranger, who became a terrible burden. She took her in knowing that she would be difficult to protect. Who was the woman and what was involved in keeping her? *She looked very Jewish, her Polish was faulty. My entire building, with all its tenants from top to bottom, knew that I was hiding this Jewess. She stayed with me not because I wanted her to, but rather against my will. No one wanted her. I could not place her. She was the aunt of my girlfriend's sister. I did not even know her. My husband and I were distressed about keeping her. She was a very limited person. I did not feel that she was the kind of person for whom we should risk our lives. I felt that I should have kept someone whom I saw as more valuable. . . . She passed as our maid. When she went out she put a bandage over her face, not to be recognized. Many friends felt that it was too dangerous to keep her. She was so stupid. At every step she would reveal her origin. I did not like her. It hurt me that instead of her I was not saving someone else.*

I interrupted Felicia, asking why indeed did she keep her if she felt so hostile about her. Taken aback, and only after a moment's hesitation, Felicia answered through her own questions. *What could I do? Could I have thrown her out? She would surely perish. The only way out she had was to go to the bridge and jump into the river. Besides, she even wanted to commit suicide. When I brought her home she was standing on the bank of the river, ready to drown herself. This was after I tried to place her in a number of homes as a maid and when all these people, one after the other, refused to keep her. . . . She felt that my place protected her fully. She did not realize that she survived because our neighbors valued us and respected us. Actually, some of them respected us and others were afraid of us because of our connection to the underground. But this was beyond her capacity to understand.*

Clearly, Felicia would have preferred to help those she cared for and, if not that, at least someone whom she defined as valuable. And yet, the need to help was so compelling that it allowed her to disregard all other considerations. She simply let this need to stand up for the helpless and dependent get the upper hand.

Just because a person is prejudiced or has strong personal likes and dislikes does not mean that the same person would not have an overall desire to help. If strong enough, such a desire can indeed overshadow prejudice. Perhaps the more threatening the situation, the greater the likelihood that prejudice will be discarded. Where the threat is severe, the victim's plight may reactivate the helper's need to be charitable. This need, in turn, may appear as an abstract force unhindered by personal likes and dislikes. In short, to the extent that a predisposition, almost a compulsion to aid others, is governed by a higher moral order, it may become independent of personal likes and dislikes. This was especially true for the anti-Semitic helpers.

What conclusions can I draw from the analysis of my findings? Throughout this chapter I have tried to demonstrate the presence and describe the meaning and interrelationships among six basic characteristics and conditions associated with the rescuers. To reiterate these characteristics and conditions, I refer to the rescuers' (1) individuality, (2) independence or self-reliance in pursuing their personal values, (3) matter-of-fact views about rescue, which come together with the insistence that there was nothing heroic or extraordinary in their protection of aiding Jews, (4) long-lasting commitment to aid the needy, a commitment that began before the war and that in the past infrequently involved Jews, (5) the unplanned and gradual beginning of rescue, at times involving a sudden even impulsive move, and (6) universalistic perceptions of the needy that overshadow all other attributes except their dependence on aid.

I began this chapter by suggesting that on a deeper, more basic, level even the social heterogeneity of these rescuers may fit into an overall theoretical explanation. Now I am prepared to show how some of these varied characteristics may express both individuality and independence. I am referring to the fact that the leftists, the intellectuals, those who were pious in a special independent way, showed a certain propensity for Jewish rescue. In wartime Poland leftist parties were unpopular and those with leftist inclinations were in a definite minority. As such, therefore, they stood out within their environment. Similarly, not to take seriously what one heard in a Sunday sermon was equivalent to religious disobedience. In a country where religion was taken very seriously and where the Catholic Church wielded supreme power, this was unusual behavior. Closely related to this was the rejection of the view that Jews were "Christ killers" and hence deserved to be punished.

Finally, too, the better educated, the intellectuals, were a definite minority. They, too, were only loosely integrated into their communities. To be an intellectual by definition implies independence and an ability to follow one's own convictions of what one believes to be right or wrong. Because each of these characteristics reflects a certain degree of separateness and independence their presence adds further support to the hypothesis that, as a group, rescuers were indeed independent individualists. Moreover, the fact that some of these rescuers were motivated by religious, political, or abstract moral principles challenges London's assertion that they were guided by parental values only. Without denying the family as a source for personal convictions and conduct, I am suggesting that the rescuers' individuality, independence, and commitment to follow personal values might have originated in religious, political, familial, or any other realms. In terms of practical results, it did not much matter whether the rescuer was a devout Catholic, a Communist, or someone who fought for justice. No matter what the starting point, the qualities that resulted all seem to have converged into a force that facilitated the rescue of Jews.

Despite their social heterogeneity, these rescuers shared a common past that involved standing up for the poor and the downtrodden. For them their history of charitable acts eased the move to Jewish rescue. Thus, to the extent that these Poles accepted and took for granted their selfless actions on behalf of others, they were able to engage in the more perilous protection of Jews.

These Poles shared a cluster of highly interrelated perceptions and attitudes about Jewish rescue. They accepted and took their aid for granted, describing it in a matter-of-fact way. Practically all rescuers defined their selfless protection of Jews as a simple duty. Portraying their aid to Jews in sincere and modest terms permitted them to strip it of its extraordinary qualities.

What does this persistent pattern mean? The sincerity and consistency with which rescuers refused to see their help as extraordinary reflect their modesty. Beyond this, what other functions do such attitudes perform?

It is possible that without stripping their protection of Jews of its heroic,

life-threatening, and extraordinary qualities, they could not have participated in these actions. For to have realistically assessed the risks inherent
in the protection of Jews, these rescuers might have become overwhelmed
and paralyzed into inaction. In short, had they focused their attention on
the ever-present possibility of death, they might have been unable to act. I
am convinced that many more Poles wanted to save Jews but did not
because they could not overcome their apprehension about death. I have
no evidence that would show the proportion of people who had a predisposition and history of aiding the needy and who wanted to protect Jews but
did not. All I have are a few scattered examples.

To recall, when talking about their decision to send away the Jewish girl,
Staniewska said that all along her father was against keeping the child.
Why? *He was a realist; he saw things more clearly and perhaps this is why
he was more afraid.* Similar conclusions are suggested by a wartime diary
of a prominent Polish doctor. In it the physician deplored the horrible
plight of the Jews, thereby showing his deep sympathy. Yet his aid to Jews
was limited to a single purchase of food. Indeed, he devotes considerable
attention to Nazi punishments for harboring Jews.[46]

In another instance a Polish teacher vigorously condemned anti-Jewish
atrocities, particularly those committed by his countrymen. He and his
wife felt deeply for the unfortunate victims. In fact, the same man said that
the greatest tragedy in his and his wife's life occurred when the Nazis had
liquidated the local ghetto. His wife, out of sheer despair and frustration,
reacted by hitting her head against the wall.

Keenly aware of and so strongly opposed to what was happening, what
did the couple do? What concrete actions did either of them take? Once,
on his way home, this Pole stopped at an inn and realized that local
peasants were preparing a raid against illegal Jews. He publicly scolded
them. So very knowledgeable about the Jewish plight, so deeply shaken by
it, this teacher limited his reactions to a purely emotional and verbal
level.[47]

Additional scattered examples exist of Poles who identified with the suffering of Jews and who would have liked to help but did not. No doubt in
some of these cases, as in those I have mentioned, the realistic appraisal of
the situation prevented these Poles from acting upon their inclinations.
Most likely in such cases extensive awareness of the situation interfered
with whatever desire to help they might have had. Thus a history of selfless
protection of the needy may be a necessary but not a sufficient cause for
saving Jews.

By definition this did not apply to the rescuers. Those who did engage in
these life-threatening actions succeeded in minimizing the possibilities of
disaster inherent in their actions, at least on the verbal and emotional
levels.

That the rescuers unlike their Jewish charges tend to minimize the importance of their help and describe it in matter-of-fact ways is understand-
able in that, for them, standing up for the helpless and the persecuted was

an extension of a well-established pattern. Helping behavior was an integral part of their life-style, which explains why they engaged in this behavior without premeditation. Some reacted not only spontaneously but at times even impulsively. So compelling was their need to help that it overshadowed personal likes and dislikes and canceled out the effects of some disagreeable attributes of the dependent and helpless. All that remained was the victim's suffering and helplessness.

Suggested by my findings is an overall theoretical construct of behavior aimed at protecting defenseless victims. I will be referring once more to this tentative theory in the next and final chapter.

CHAPTER 11

Conclusions

In wartime Poland a move to the forbidden Christian world gave Jews a chance to live. To survive in such a menacingly dangerous environment, one had to steel himself or herself to danger and hope that luck would be on one's side. If one wanted to survive, there was no choice but to try. The alternative was almost certain death.

Aware how precarious their predicament was, Jews on the Christian side waged a fierce struggle. In this struggle some found support and aid from Poles who, by the very nature of this association, put their own lives at risk as well.

While for Jews, failure to locate such helping Poles effectively eliminated their chances of survival, the opposite was true for these protectors; refusal to shield Jews would have removed a serious threat to their lives.

Brought together by different circumstances, the rescuers and the rescued, the Poles and the Jews, eventually came to face similar and highly interrelated perils. Considering the life-threatening nature of Jewish rescue, who were the Polish rescuers and what motivated them to risk their lives for the persecuted Jews?

Not unexpectedly, the Holocaust literature deals with these issues. Relying mainly on speculation, personal observations, personal experiences, as well as scattered case histories, previous writings on righteous aid fall into one of two categories. The first assumes that righteous Christians cannot be identified by special characteristics. Those who subscribe to this view argue that for a Jew it was impossible to predict who would and who would not help. There is some validity to this assertion. The literature and my research contain many illustrations of help extended by the least likely individuals, and denied by those who promised it, or who, because of a special relationship of love or friendship, were expected to provide it.

Those who feel that rescuers have no special characteristics in common also point to cases of known anti-Semites who, during the war, risked their lives for Jews. They argue that since such unlikely candidates as avid anti-Semites became Jewish saviors, it is futile to expect to find special characteristics common to all.

Key then to this first position is the presence of anti-Semites among those who rescued. However, in my own research on avid anti-Semite helpers the same few names seem to recur over again. Therefore, the inevitable conclusion is that they made up a small discrete group. In addition to being small in number, they have other aspects in common. They are all socially prominent, devout Catholics who were politically active. These rescuers also had long histories of charitable acts on behalf of the poor and needy. The guiding force behind their good deeds was their religious convictions.

During the war this small group of anti-Semites became appalled by Hitler's inhumanity to the Jews. As introspective Catholics and intellectuals, they felt that their own anti-Semitism might have indirectly contributed to Jewish destruction. At this point, for the first time, they perceived the Jews as in need of help. Their traditionally antagonistic image of Jews became overshadowed by a new image of Jewish helplessness and extreme need. Since for these Poles helping the needy was an established pattern, the plight of the Jews demanded their attention. For them, then, aid to Jews satisfied different needs. It alleviated the guilt they might have experienced because of a belief that their own anti-Semitism had contributed to Jewish destruction. Furthermore, as patriots, saving Jews was part of opposing the Nazis. But such help by overt anti-Semites did not signal an elimination of anti-Jewish prejudices. Instead, it signified an armistice, a lull in hostilities. Rare and special as rescue by overt anti-Semites was, it belongs to an intriguing exception, and as such does not deny the possibility that most rescuers may share a special set of characteristics and motivations.

In fact, the second position accepts as a starting premise the idea that righteous Christians have certain characteristics and motivations in common. However, these shared characteristics and motivations are neither uniform or consistent. Not surprisingly, neither are they presented as a set of consistent explanations. What is the nature of these hypothesized characteristics and motivations?

Class affiliation and political involvement have been suggested as conditions that predisposed people toward Jewish protection. Some have argued that lower class individuals, because of their own disadvantages, could more readily identify with the suffering of Jews and hence were more likely to help them. Others felt that lack of contact between peasants and Jews and the resulting estrangement made the former less rather than more responsive to the Jewish plight than were other segments of the population.

Some have assumed that intellectuals, more aware of the grave implications of Nazi success, were more willing to protect Jews.

All agreed that, because of direct competition with Jews, the Polish middle class was unsympathetic to their plight and hence least likely to save them.

Relying on Polish rescuers themselves and those indirectly described by the survivors, I have examined each of these assumptions. This I did by dividing each group in terms of class and by comparing the rescuers' class distribution to the class distribution in Poland for that period. Results from such comparisons indicate that, contrary to what some have expected, lower class individuals show no special propensity for Jewish rescue. The proportion of lower class helpers and the proportion of lower class individuals within the country as a whole were identical. In contrast, the prediction that peasants were less likely to help Jews than others gains support. Fewer peasants became rescuers than their numbers in the general population should have warranted. Even though proportionately peasants were less inclined toward Jewish rescue than other groups, a substantial number of them did participate in selfless protection of Jews.

As expected, the intellectuals were more prone to Jewish rescue than any other segment of the population.[1] Unsupported by my evidence was the assumption that the middle class would refrain from protecting Jews. In fact, proportionately, there were as many middle class persons among rescuers as in the population at large. Class affiliation, then, was only weakly related to Jewish protection; thus it qualified only as a partial predictor of people's willingness to risk their lives for others.

Political involvement per se does not seem to be related to Jewish rescue. The majority of these rescuers were politically inactive. Only among the few politically involved helpers could a strong positive association between leftist politics and Jewish rescue be seen. Although leftist inclinations did faciliate help to Jews, they in no way accounted for the bulk of Jewish rescue because most rescuers were politically uninvolved.

Among other conditions held responsible for saving Jews were religious beliefs and values. Religious beliefs and values, however, were far from uniform. The formal teachings of the Church could be interpreted in various ways even by the clergy. Lay Catholics, too, could interpret their religious obligations independently and differently from both how the Church and the clergy interpreted them.

The traditionally anti-Semitic Polish Catholic Church had no uniform wartime policy regarding Jewish extermination. Absence of an official posture left much latitude for clergy and the lay public. As a result the clergy responded in a variety of ways, ranging from denouncements to self-sacrifice. No adequate comparisons of the proportion of clergy among helpers and their proportion in the population at large were possible, but limited analysis revealed no major role for the Catholic clergy in Jewish rescue and led me to the tentative conclusion that their participation in the rescue of Jews was no more extensive than that of most other segments of the population.

Setting aside the extent of the clergy's participation, it is agreed that they, more so than others, concentrated on saving Jewish children. Most of these children, who were moved to convents and monasteries, were baptized; this practice led to the accusation that in saving Jewish children the basic motivation of the Church was to gain converts. This has been disputed by the pious saviors, by some Jews, and by some available facts. From a number of cases it is clear that conversion of Jewish children was undertaken cautiously and often only after special permission was granted by parents or guardians. Moreover, Catholics close to the process explained that, after baptism, Jewish children were more likely to feel Christian and therefore had a better chance of avoiding giving themselves and their rescuers away. Most Jewish survivors I interviewed who were children at the time were convinced that, whether they were formally baptized or not, they derived much comfort from the Catholic religion. Still, it is impossible to settle the question of the clergy's primary motivation.

Apart from the role of the Catholic clergy itself, what impact did religious values and beliefs have on the rescuing of Jews? To what extent were righteous Christians prompted to help by their religious convictions?

Information about the religion of many helpers was missing. Only a minority of the Poles I interviewed directly were deeply religious; these few credited their religious convictions with prompting them to initiate and sustain Jewish rescue. Yet none of these helpers were devout or blind followers of the official doctrines. Their brand of religiosity demanded of them high humanitarian standards, which they followed regardless of external pressures. Indeed, when and if their religiosity was in conflict with the Church's pronouncements, their own brand won. Rare as this form of religiosity was, it could and did lead to rescue.

On the other side, scattered cases in the literature show that at times religion was used as a justification for Jewish destruction. Some members of the clergy were known to urge their parishioners to support Nazi policies of Jewish extermination. In a few cases the rescuers themselves were subjected to religious anti-Jewish pressures. Whereas a few pious helpers in my study did vacillate under the clergy's pressure to desist from saving Jews, none succumbed, and the majority had no trouble resisting this pressure.

Conventional religion was flexible. In the absence of a uniform official position toward the Nazi destruction of Jews, it could lead to either saving or destruction. Therefore, religious conviction per se cannot be held responsible for Jewish protection: only a special kind can.

When trying to account for the rescuing of Jews some have pointed to friendship. Friendship does require some sacrifices: therefore, the assumption that it would lead to Jewish protection seems plausible enough. On the other hand, given the life-threatening quality of Jewish rescue, is it realistic to expect friendship to account for such behavior?

There were many instances where Poles denied help to their Jewish friends. Some Jews turned away by friends received aid from total strangers. A minority of the Jewish survivors in my study (12 percent) were actually denounced or blackmailed by their Polish friends. Among the rescuers themselves, only a minority spoke of friendship as a motivation for Jewish rescue, and only 9 percent of these Christian helpers limited their protection to friends alone. In 65 percent of cases their aid involved both friends and strangers.

More extensive information from my interviews showed that at times circumstances beyond their control prevented some rescuers from protecting friends, even though they would have liked to. In most cases, however, an inclination to aid friends did not exclude help to others. Since most rescuers risked their lives for those who were not friends, friendship cannot be held responsible for Jewish aid. Furthermore, whereas friendship did not necessarily lead to rescue, strangers and mere acquaintances more often than not were the recipients of selfless aid. In short, those who did rescue would quite naturally have preferred to save friends, but even among these people a need to rescue had to come first. Friendship with Jews alone was not enough to make rescuers out of those who were not otherwise inclined.

What conclusions were suggested by these different yet commonly held assumptions? As unexpected as they were rare, anti-Semitic rescuers were more visible, leading to the mistaken assumption that many Polish helpers were overt anti-Semites. This is not so. For even though most rescuers were imbued to some degree with their society's anti-Semitism, only a few fit the definition of active anti-Semites. In fact, much more tolerant than other Poles, the rescuers came from different walks of life, related differently to politics and religion, and only rarely confined their aid to friends. While each of these conventional categories bears some association to Jewish rescue, none qualifies as a basis for classifying these helpers or as an explanation of their participation in this life-threatening behavior.

From the isolated attempts to study Jewish rescue systematically, I borrowed the concept of altruism and individuality. Risking their lives for others regardless of sacrifices and in opposition to societal expectations, these rescuers fit the definition of autonomous altruists. A close-range analysis of these selfless helpers yielded a cluster of shared characteristics and conditions. Highly interrelated, common traits refer to the rescuers': (1) individuality or separateness that resembles closely London's idea of marginality; (2) independence or self-reliance to act in accordance with personal convictions, regardless of how these are viewed by others; (3) broad and long-lasting commitment to stand up for the helpless and needy; (4) the tendency to perceive aid to Jews in a matter-of-fact unassuming way, which comes together with consistently strong denials of any heroic or extraordinary qualities; (5) an unpremeditated, unplanned beginning of Jewish rescue, which could either have happened gradually or suddenly, even impulsively; and (6) universalistic perceptions of Jews that defined them as helpless beings and as totally dependent on the protection of others. Closely connected to this is the ability to disregard all attributes except those that expressed extreme suffering and need.

With only a few exceptions, but in different ways and in different degrees, the Poles in my study did not fit into their milieux. Some were conscious of their differences while others did not mention them and might not have been aware of them. Whether they were or were not aware of their individuality, similar consequences follow. What are these consequences?

Those who are on the periphery of their community are not strongly controlled by it. Almost by definition those who stand out and do not blend within their environment are freer than those who are well integrated.

Less constraint and more freedom of action imply a certain degree of independence. For example, being leftist or unconventionally religious in wartime Poland required a certain degree of individuality. As important, an overwhelming majority of rescuers were not prepared to abandon positions because others might then view them differently. They were relatively free from external pressures, and others perceived them as such. Independence goes hand in hand with strength and freedom.

The relative independence, strength, and freedom of these rescuers suggest that they were able to act in accordance with their personal moral imperatives, which involved a strong desire, almost a compulsion, to stand up for the needy, the persecuted, and the downtrodden.

The imperative to help could and did come from various sources: some helpers said they had been influenced by religious teachings, others by political arguments, still others traced the source of their unique moral values to family life. No matter what the source, these values seem to have become ingrained in the individual. All experienced them as powerful and compelling guides to personal conduct. In addition, such imperatives appear to have been incorporated into the moral makeup of these individuals well before the war. In case after case, there is a long history of giving aid to those in need.

Most important, included in these imperatives was the clear direction that help was to be offered to anyone in need, regardless of who they were. The fact that everything except the dependence of those who were aided seems to have faded into the background suggests that these helpers saw their obligations in the broadest terms—in applying the idea of man's obligations to help his fellow man. Moreover, the strength of this moral imperative, together with its universal quality, may in part explain why these Poles were willing to risk their lives for strangers and why they insisted that they did not see Jews as Jews but only as haunted, persecuted human beings in desperate need of aid. The existence of this universal quality of help is also supported by the fact that at times these Poles saved Jews for whom they had neither respect nor liking.

Because help to Jews was an extension of the rescuers' past protection of the needy, they experienced and defined it as a natural and obvious reaction to the sufferings of others, as a mere duty that had to be performed.

Perception of this life-threatening behavior as a duty was also closely related to the way in which the help was extended. Such aid began either as an automatic, impulsive reaction or in a gradual but unpremeditated way. In none of the cases was it a conscious decision in which its possible implications were systematically considered.[2]

Inasmuch as these Christians perceive their actions as duties that had to be performed, as natural and obvious, it is not surprising that they also tend to describe them in a matter-of-fact way. Moreover, in line with such perceptions is their insistence that their rescue of Jews was neither extraordinary nor heroic.

Most of the Poles I interviewed insisted that the term heroism does not fit them at all. To support their position they pointed to their fears and anxieties: while many managed to push such fears into the background, none denied that they existed.

Another way these rescuers minimized the heroism of their actions was by emphasizing that the dangers emanating from Jewish rescue were only a part of an overall threatening environment. They pointed out that people could and did die for nothing at all. They argued that, since life was full of constant and unexpected perils, helping Jews was just one additional reason for dying. Compared to the ever-present threats, aid to Jews was not as dangerous and hence not as extraordinary as some tend to see it.

To be sure, although much of the Nazi terror took place for no apparent reason, it is nevertheless true that by helping Jews Poles were courting personal tragedy. Aiding Jews was an invitation for disaster.

The fact that the rescuers viewed their rescuing of Jews as a natural reaction to human suffering can, in part, be explained by their strong moral imperative to help the needy, an imperative that was a well-established tradition. Besides, the rescuers' modest matter-of-fact attitudes support the idea that people tend to accept and take for granted their own patterns of behavior no matter how extraordinary they may seem to outside observers.

Finally, also, to participate in life-threatening and extraordinary actions, people may need to redefine such activities and see them in less threatening ways. For example, those who go on extremely dangerous missions, to be effective, must at least push the inherent threats into the background. Similarly, people who live under uncertain and potentially explosive conditions do not dwell on these negative aspects and do not allow these negative forces to overpower them.

Fully aware of what punishment was attached to the rescuing of Jews, these helpers refused to focus on it. Some admitted that concentrating on the existing threats would have paralyzed them into inaction. Refusal to see their actions in anything but a matter-of-fact way was one of the conditions facilitating aid to Jews. No doubt, too, an unspecified number of Poles would have liked to participate in Jewish rescue but did not. Such Poles, I believe, unlike the rescuers, were unable to ignore the existing perils. Instead they allowed themselves to be overcome by them.

Calling for much further exploration and analysis, the theoretical formulations that grew out of my research suggest the following set of interrelated hypotheses.

The less integrated into a community people are, the less constrained and controlled they are by the community's norms and values. Thus freed of constraints people are more likely to resist the pressures of the community and act independently.

A high level of independence, then, implies a greater amount of strength and freedom to act in accordance with personal inclinations and values. The righteous Poles' personal inclinations had to do with moral imperatives, which were expressed as a strong desire to help the needy. The wish to stand up for the downtrodden was a powerful, compelling force, capable of overshadowing the particulars of the needy, except those which had to do with the dependence on aid. At times this wish to help others could and did overcome even personal dislikes and prejudices.

The longer people act in accordance with such strong moral imperatives, the more likely are these actions and values to become traditional patterns. The more firmly established such actions become, the easier they are to follow and the greater the likelihood that they will be taken for granted and defined as natural reactions and as a duty.

If such actions become well established and are viewed in a matter-of-fact way, there will be a greater likelihood to deny the extraordinary or heroic qualities of these acts.

Arrived mainly through the inductive method, these principles inevitably raise a number of questions. One of these has to do with the relative importance of the variables under consideration. What relative impact do they have upon self-sacrificing behavior in general and the one under study in particular? Another touches on the exhaustiveness of these explanations. That is, there might be additional or alternative interpretations that I have neither discussed nor explored. Still another issue closely related to all others has to do with the overall validity of my interpretations.

To be sure, each of these issues is more easily raised than answered. Still, it is tempting to venture a few tentative guesses. This is all I can do.

Concerning the relative significance of the factors examined, it would seem that the tradition of standing up for the needy is most significant. Next seems to be individuality. The rest of the characteristics and conditions tend to follow from these two basic factors. Whereas the many different conditions apply to the Polish rescuers in varying degrees, the need to help is apparent in all cases. Similarly, only in a few exceptional cases was individuality absent. For these exceptions, alternative explanations might be more appropriate. I have applied alternative yet related explanations to the discussion of anti-Semitic rescuers. To recall, of the Poles I interviewed directly two, Dunski and Horska, were not individualistic; both fit well into their environment. Both were overt anti-Semites, in the sense that before the war they had participated in anti-Jewish actions and to this day have expressed anti-Jewish views freely.

While both were well integrated into their community, they did exhibit a high degree of independence, and traditionally have had a compulsion to stand up for the needy. In fact, these two and all other anti-Semitic rescuers shared a few special characteristics. They were socially prominent, deeply religious, and nationalistic. All overt anti-Semitic helpers wanted to rid Poland of Jews, but not by murdering them. During the war Jewish suffering and destruction appalled them. They felt that their own anti-Semitism might have contributed to this destruction. To atone for their past sins and to alleviate their guilt, they rescued Jews. They seemed to be particularly introspective people, with a keen sense of justice. Still their risk-taking behavior on behalf of Jews did not eliminate or change their anti-Semitism. To be sure, among rescuers such overt anti-Semites were very rare. As an interesting exception, however, they suggest that individuality can be absent from some rescuers. All the other explanatory conditions—the moral compulsion to help the needy, the matter-of-fact view about rescue, and the closely related tendency to diminish its importance—were present both among the few anti-Semitic helpers and all others.

In a sense, the chapters on class and politics, religion, friendship, paid helpers, and anti-Semites each deal with alternative explanations. As previously suggested, intellectualism, unconventional religious views, and leftist inclinations all imply a certain degree of individuality and independence. Therefore, rather than denying the validity of my explanations, these chapters tend to support them.

Turning more directly to the issue of validity, can we establish that the conditions and characteristics thus far presented are peculiar to selfless rescuers only? Resolving this problem properly would have required comparing these helpers to a control group of those who refused to aid Jews. For practical reasons this was not possible. In the absence of a real control group, I relied on an approximation and used Poles who saved Jews for payment.

Information about those who saved for money comes indirectly from accounts of Jewish survivors who received such help. These paid helpers, unlike the rest, were motivated by the payment they would receive for the services they rendered and without the financial incentive they would not have helped. As a group they do not fit the definition of rescuers. Whenever possible, I tried to see if they also differed from the rescuers in other ways besides their money motivations.

Direct information from my interviews showed that only a few of the paid helpers were individualistic. While a substantial proportion of them were independent, none were guided by moral imperatives that required them to stand up for the needy. Whether propelled by hunger or by greed, the paid helpers were motivated by the desire for tangible rewards; their commitment to the protection of Jews was weak and could be easily terminated by external threats. In fact, unlike the rescuers, most of those who helped Jews for money regretted their initial decision. Most would have liked to have terminated the association, and many in fact discontinued their aid. Also, these Poles mistreated their charges, by starving them, demanding more money, and sometimes threatening them with denouncements and death. No such harassments were attributed to the rescuers.

Comparisons of Polish rescuers and those who saved Jews for money accentuate the difference between the two groups. This in turn supports the overall validity of my findings.

The differences between the paid helpers and rescuers lead to other intriguing conclusions. Namely, they show the limitation of money as an incentive. Once their desire for money was fulfilled, the paid helpers saw no reason for risking their lives. Reality, with its threats and dangers, began to reassert itself. But because the Polish-Jewish relationship could not be terminated without serious risks, it continued. Inevitably the paid helpers began to feel trapped and helpless, and as they did, they began to vent their frustrations on the Jews whom they saw as the source of their fears and apprehensions. In short, money is but a transient unstable incentive for life-threatening behavior.

Unlike these paid helpers, those who risked their lives for others for no tangible rewards were independent individualists who preceived aid to the needy as a simple duty. And even though they refused for special reasons to allow their acts to be described as heroism, they were true heroes.

The unprecedented cruelty and devastation of the Holocaust led to extreme evil and less often to extreme goodness, extreme goodness that is epitomized by the rescuers. Still, were it not for the Holocaust, most of these helpers might have continued on their independent paths, some pursuing charitable actions, some leading simple, unobtrusive lives. They were the dormant heroes, often indistinguishable from those around them.

These rescuers acted in ways that were natural to them—spontaneously they were able to strike out against the horrors of their times. The very presence of such people must give us hope.

Postscript on Methodology

Much of the published material on the Holocaust is based on personal accounts, casual observations, and scattered case histories. Important as this material is, it was collected more for the purpose of showing the tremendous tragedy of the Holocaust and not as a basis for systematic study. One could only draw a few, very tentative generalizations from these accounts about who had helped and why they had done so, and I soon became convinced that someone needed to approach this subject more systematically.

But systematic study called for more clearly spelled out aims and procedures. I decided that my interest in the rescued-rescuer relationship necessitated examining each of the partners. But who should be included in each group, and where would I find them? Following my personal inclination, I decided to focus on Poland.

I knew that most Polish Jews who survived the war did so in concentration camps, as partisan fighters, and by passing as Christians. In reality these different modes of survival could become blurred, because fugitives found it necessary to switch from one coping strategy to another in strange and unexpected ways. To maintain my special focus of interest, I excluded from my study those Jews who never lived illegally among non-Jews and those who survived mainly as partisan fighters. In the end, however, because Jews were forced to live under a variety of circumstances, the group studied includes some who, in addition to "passing," had other experiences as well.[1]

Choosing Polish rescuers was simpler. Any Pole who knowingly aided Jews would fit my criteria. But here too a process of self-selection was at work. Helpers who performed single charitable acts were rarely, if ever, identified by others or by themselves as rescuers. For example, when Jews mentioned Poles who had shown them how to reach a safe place, they limited their description to this particular event. Thus, lack of information would in turn prevent such Poles from becoming a part of my sample.

Initially disappointed by the literature, I then returned to it with new expectations and demands. I considered more carefully general publications on Polish-Jewish relations before and during World War II. Not only did these publications provide me with a historical setting for my research, but they also broadened my own views on the subject.

My search for insights into the rescued-rescuer relationship led me to realize that, although limited, the available publications could serve well as general guides.[2] Also, the absence of overall explanations, combined with inconsistencies, encouraged my receptivity to new ideas. Without a model to lean on, and without precise expectations to follow, I had to explore and to improvise.

At first, because published accounts and case histories contained only descriptions and no explanations, I found them disappointing. Later on, while collecting facts that would lead to explanations, I became aware of the usefulness of these materials. Many of these were collected by different historical commissions. Others were written at the authors' requests. Still others came from published sources. Each case is identified by name, and all the materials are well documented. I felt that these accounts and case histories could become a part of the sample of Poles and Jews.[3] But as I proceeded to extract from them basic facts, I realized that only a fraction contained adequate information. The majority therefore had to be discarded.[4]

Another part of the Holocaust literature contains memoirs of Poles who saved Jews and of Jews who survived by passing. Compared to the brief accounts and case histories, these memoirs are richer in factual material. Jewish memoirs also include varied descriptions of rescuers. I recorded the relevant information about these helpers. Later on each was treated as an additional but indirect case. And yet, memoirs have also a variety of limitations. Because writers tend to be educated they represent a special group. Such special characteristics inevitably limit the groups' representativeness.[5] Another shortcoming pertains to non-Jewish authors. With few exceptions, Polish memoir writers were members of RPZ (Council for Aid to Jews). When writing about their war experiences these Poles focused on the activities of the organization, giving the impression that their role in protecting Jews was an organizational matter rather than a personal one. Descriptions of personal actions, motives, and reactions are noticeably absent in these memoirs.

When dealing with events that happened so long ago, time inevitably becomes an issue. One commonsense solution would be to use information collected as soon as possible after the events. Accounts and memoirs written during the war or right after fulfill this requirement. Yet, not wishing to neglect important facts that only became available recently, I used both early and late evidence.

Closely related to the study of the past is the issue of accuracy. Did the described events really happen? Are the reports valid? Occasionally, I found two or three people who independently referred to the same events. Comparison of information from such diverse sources revealed no serious differences. They seemed to reflect different perspectives rather than out-right distortions. The same seemed to be true for evidence collected at different times from a single individual. (Later I will discuss more fully this issue of validity.)

Looking for a wider coverage and for less specialized groups, I moved from published accounts, case histories, and memoirs, to archival collections. After the war in Poland and in other European countries, historical commissions were established for the purpose of preserving materials about the Holocaust. Through these commissions thousands of eyewitness testimonies were collected, most of them from Jewish survivors. Included among them are the stories of Jews who were in concentration camps, who spent the war in Russia, or who were in partisan units. Only a small proportion deals with those who tried to survive on the Christian side, and an even smaller proportion with Poles who were protecting them.

I decided to examine three different archival collections: at YIVO, the Institute for Jewish Research in New York City, at the Jewish Historical Institute in Warsaw, and at Yad Vashem in Jerusalem and Givataim. Access to these collections depended on the particular order in which the materials were organized.

At YIVO most testimonies are ordered by geographic regions, a classification that does not bear on my subject. Therefore, for fewer than a hundred cases I had to pore over more than 2000 eyewitness accounts. In Poland I was more fortunate. The Jewish Historical Institute in Warsaw has one of the most extensive collections. At the time of my visit they were in the process of rearranging the materials, but much of it was already ordered in terms of the issues I studied, and I had easy access to the cases I needed. Finally the extensive collection at Yad Vashem is organized in terms of a variety of categories, some of which referred to my special interest which, in turn, saved me much unnecessary searching.

Still none of these archives contain uniform data. When looking at any of these testimonies it seems clear that whoever was collecting the evidence followed no special guidelines. As a result, these eyewitness accounts cover a wide and scattered array of topics. Many lack elementary facts, while others report in detail about experiences which are irrelevant to my special concerns. Incompleteness, plus the particular kind of information forced me to exclude most cases. In Poland, for example, of the many hundreds of cases I examined, all but seventy-four had to be discarded because they lacked basic information.

Moreover testimonies in these different archives were collected at different times, some right after the war, while others only a few years ago. The earlier cases are usually less complete than the late ones. Perhaps they were collected in haste. Perhaps those who gathered this information, because of lack of experience, were not aware of the value of different kinds of facts. Whatever the reasons, the early testimonies at YIVO and the Jewish Historical Institute contain less extensive data than some of the late accounts collected by Yad Vashem.

More detailed information, however, does not necessarily guarantee its usefulness. As with published accounts and memoirs, here too I used both the early and later testimonies if they contained the information I needed.

But what kind of information was I looking for?

As is traditional for a sociological study, I wanted to know about the usual background characteristics of each group. In addition, I tried to collect evidence about attributes peculiar to the wartime situation and to the special subject matter. For example, I tried to ascertain the physical appearance of Jews who had passed for Christians, whether they were assimilated, about circumstances surrounding rescue, kinds of help, how and in what way it was offered.

My sociological training made me especially aware of social integration. More specifically, I wanted to know how these helpers fit into their environments. How did the rescuers and rescued feel about each other and about the most crucial part of their relationship—help. Finally, since all Polish–Jewish associations are overshadowed by anti-Semitism, I had to know how anti-Semitism fit into this relationship.

Without a general theory to test, I wanted more data than any of my sources could offer. Important information that was conspicuously absent from both the published and unpublished sources had to do with motivations, attitudes, values, and meaningful descriptions of the relations between the rescuer and the rescued. Especially disturbing to me, however, was the absence of systematic information about anti-Semitism and how it applied to the Christian helpers.[6]

Absence of valuable data, more so than any other consideration, convinced me that I would have to conduct my own in-depth interviews. As a Holocaust survivor who for over thirty years had shied away from any mention of the war, I knew that I would have difficulty finding suitable Jewish cases. As for the rescuers, I had no way of knowing whether they would talk to me or not. I remembered Poles as avid anti-Semites, and I feared that they would refuse to become a part of the study. Yet I shared common experiences with both of these groups. Those of us who survived the war lived under similar circumstances and spoke the same language. There are even special expressions familiar only to those who had something to do with passing and with the protection of Jews.[7]

I decided to start by interviewing Jewish survivors who lived in the United States. I felt that it would be easier to make necessary adjustments here than in a faraway country, where I could ill afford to lose a case because of poor wording or other shortcomings.

I expected the survivors to be reluctant, but I was hopeful that my personal experience would help. For my first case, a friend gave me the phone number of his close friend. As noted earlier, this first attempt met with a flat rejection: "My past is my own business, it is too private to be used by anyone . . ." Later on I became more cautious. The same friend who directed me to the first contact arranged a meeting with Szymon Rubin, a science professor[*] at a leading university. This time, however, I called only after Szymon had agreed to meet me. At the end of this first interview, I asked whether he could help me arrange an interview with another survivor, to which he readily agreed. However, the person he had in mind, Renata Shein, was less willing. Fortunately, Szymon insisted and kept calling her. Eventually she agreed to talk to me on the phone, still without promising an interview. Only after I spoke to her did she agree to meet me.

Very sensitive, Renata found it hard to talk about her past. For two years she had lived with her father in a low-ceilinged attic in which standing up was impossible. She suffered continual hunger and thirst, and washing was an unheard-of luxury. She was protected by two strange Poles, a father and his forty-year-old unmarried daughter. The father was a janitor, while the daughter worked as a fortune teller. These two Poles were fearful and reluctant to continue protecting Jews, which only added to the fugitives' hardships.

During the interview Renata Shein became relaxed and cooperative. Not only did she find me two more subjects, but she directed me to her relatives in Warsaw who also proved extremely cooperative.

From then on, I followed the same procedure. Someone whom each survivor knew well would call first to explain the nature and reason for my work, and the fact that I too was a Holocaust survivor. Only after the person agreed to have me call did I follow up with a phone call. This way each of my contacts resulted in an interview.

Sensitive to the issue of wider representation, I knew that all those who would directly or indirectly come through Szymon Rubin would be highly educated. To avoid social homogeneity, after four cases I contacted Lola Freud, a friend of mine who I knew was different and who also survived by passing. Lola and her successful businessman husband had little formal schooling. Eventually I was able to include not only Lola in my sample but also a few of her friends—all very prosperous, all with little education. Around the same time, by chance, I met a young musicologist who casually mentioned that her mother had survived the war as a Christian. With some coaxing I got the mother to meet me. The people she found me were lower middle class, neither rich nor educated.

Finally, I also had a friend in Canada, Fela Steinberg, with whom I had shared an apartment right after the war. Despite our close friendship we knew little about each other's wartime experiences. The one thing we knew was that we both had survived by passing. For years my friend Fela had been inviting me to come for a visit, so in 1978 when I asked her to become a part of my study and to set up some additional interviews for me, she was glad to accommodate. As a result I incorporated five Canadian Jews into my research. One of them lived in a luxurious mansion, two in inexpensive developments, and one in a house that fit somewhere in between. None of them were professionals, none had much formal education; yet they differed in their life-styles and interests.

Of the sixty-five interviews, thirty-one took place in Poland, where I was also able to establish a variety of contacts. Of special value was the assistance of Jan Krupka, the chief archivist of the Jewish Historical Institute in Warsaw. From his archives he selected appropriate cases and called the prospective interviewees to set up appointments for me at the institute. The individuals he reached were a heterogeneous group.[9]

In addition to the institute, I had contact with a Polish journalist–historian, Władysław Bartoszewski, who has written extensively on the subject of Polish–Jewish relations during the war. Out of the country during my visit to Poland, he left instructions with Teresa Preker, also a journalist and historian, to arrange for me interviews with Poles who had rescued Jews. In most instances these were individuals whose stories appeared either in Bartoszewski's books or in other publications.

Also Renata Shein, the survivor I interviewed in the United States, directed me to her Warsaw relatives who arranged a few interviews for me. Finally, in Poland as elsewhere, before leaving a session I asked the particular individual to find me another contact. Most of them did. A few of these lived in Poland, others in Israel and the United States. I interviewed twenty-eight Christians in Poland and three in Israel. In this group too I contacted people whose social backgrounds were dissimilar: diplomats, writers, businessmen, governesses, the highly educated and uneducated. One rescuer, Maria Baluszko, a Yad Vashem recipient who saved five Jews and sporadically helped many others, when asked about her schooling simply replied: "None." After a brief pause, she added, "But my father went to school for three years. Be sure to put this down."[10]

Regardless of where a meeting took place or who the person was, I tried to follow the same procedures. Upon meeting I assured each individual that their answers would be anonymous and that the final results would not be sensationalized. Invariably I was told that were I not a Holocaust survivor they would not have agreed to talk to me. With the exception of two cases, none of the Jewish survivors had ever been interviewed about their past. All were visibly affected by the questions. Many cried. My friend Lola Freud, even as she agreed to the interview, told me that she does not like to dwell on her past because it hurts too much. When she faced me, she seemed to hesitate. She told me that her husband asked her why she was opening those old wounds. Then, rather abruptly, as if plunging into it, she said: "Let's start!"

At fifteen Lola was the youngest in a family with four brothers. Her father was a highly respected religious man who, as a well-to-do businessman, liked to help the needy. During the war he generously supplied food to Jewish prisoners of war who were forced by the Nazis to build a nearby concentration camp. To repay her father's kindness, these young men offered to place one of his children on the Christian side. This involved making false papers, smuggling the person out of the ghetto, and then finding shelter. Lola's father had to decide which of his children it should be. Because Lola had the most Polish appearance, was fluent in Polish, and as a result might have a better chance to make it, her father selected her. Preparations were begun without Lola's knowledge. Only when all was ready did her father break the news to her. Lola remembered:

At first when he explained that he chose me to live I did not understand, maybe I did not want to. . . . He repeated it, and then said, "Be a good girl, listen to them. . . . Whatever they tell you, you must obey, and I am sure you will live through the war." I looked at him and I still could not believe it, then I got very angry and said, "How can you send me away from home? Whatever is done and whatever happens to all of you will happen to me." So he said, "No, I want one of you to survive, we cannot give in to the Germans." I still did not want to go. I was very attached to my family. So he said: "If you are not going you are doing something terrible to us." And he said, "I will give you my blessing and I am sure you will be the one to tell the story of what has happened." Without giving me a chance to answer he called in the man that was to take me away.

Lola's body shook and her sobs interrupted the story. When she calmed down she told me that she had left the same day and never saw her father again.

Not all of my respondents who became emotional cried. One person, for example, developed stomach aches, while another continued only after she had taken a drink. But as the interview progressed, they felt better. At the end each told me that they were glad to have had the opportunity to talk. Without exception, each was eager to direct me to another survivor.

The Poles too were most accommodating and eager to talk. They felt that they were unjustly defined as a nation of ruthless anti-Semites and hoped that my work would help correct this image. Though willing to talk, they too were emotional about their wartime experiences.

Tomasz Jursky is now an engineer. During the war, as a teenage member of the underground, he directed Jewish strangers to his father's house. Fourteen stayed until the end of the war and survived. When Tomasz described his reunion with one of the Jews he saved, he cried. Eva Aniel-ska, a prominent underground figure who was responsible for the rescue of hundreds of children and many adults, had a hard time suppressing her sobs as she told about her best friend who perished.

Dr. Adam Estowski, who illegally extended medical help to Jews and who kept Jewish children in his house, developed chest pains during our conversation. Having recently recovered from a heart attack, he had to lie down. His wife, who had cooperated closely with him throughout the war, finished the interview. When faced with such emotional reactions I stopped my questions, kept quiet, and waited. They seemed to welcome the pause and my silence.

Before starting I expected that the interview material would give me information unavailable in any of the other sources. Because I was also interested in the quality—particularly the uniformity—of information, I devised uniform questions. When devising these questions, I was guided by the literature, by my own experiences as a Holocaust survivor, and by what I missed when looking at the brief published accounts, memoirs, and testimonies.

The initial interview helped to sharpen some of my questions. At the end of this first meeting, I asked Dr. Szymon Rubin if he would care to make comments. He suggested that I include questions about events after the war, and after the Jews and Poles parted company. Because of his suggestion, I collected some very interesting material.

I knew that Poles would be touchy about Polish anti-Semitism. Indeed, I believe that this sensitivity is partly responsible for the fact that most Polish war memoirs and testimonies are silent about the subject. Therefore, to tap this sensitive topic I posed questions that were somewhat indirect and placed them toward the end of the interview when the individual was more likely to be relaxed. For example, one question asked: "We all know that anti-Semitism has existed for centuries in different parts of the world and in different countries. How would you explain its existence? Why do you think it exists?" Another question asked: "I would like you to comment in general on the way Jews felt and feel about Poles and Poles about Jews. How do they see each other?" Another question dealt with the effect which the disappearance of Polish Jews had on Poland. Still another inquired whether and to what extent Jews who were protected by the respondent differed from Jews in general.

These questions proved useful. As if to accommodate me, the rescuers answered by giving their views about anti-Semitism in Poland, by telling how they themselves felt about Jews, and by showing the extent to which they could or could not escape the prevailing anti-Jewish ideology. The results of these questions are discussed in a number of chapters.

Because Poles and Jews had different experiences I developed a separate schedule for each. Most of the questions in these schedules were unstructured, offering a great deal of flexibility. I felt that when inquiring about personal and highly emotional topics too much structure might be offensive. In addition, open questions offer the chance to collect new and unexpected evidence and hence may lead to exciting new insights. Each individual was also given the opportunity to use a language he or she felt most comfortable with. Of the 65 people I interviewed, only five chose to speak English. They are all Jewish. But even in those five cases, the English was often interspersed with Yiddish and Polish expressions. All the rest spoke Polish. Later on as I transcribed each interview, I translated it into English.*

To preserve all I heard I decided to use a tape recorder and was pleasantly surprised by the lack of opposition. Especially in Poland I had anticipated that people would be reluctant to speak freely. But, in Poland only one person, a Jewish survivor, refused to be taped. Fortunately, he had a tendency to repeat himself. Therefore, I doubt that crucial information was

---

*Except for a single document, I have done all the translations for this book.

lost. The same man had written a book of memoirs right after the war. I read the book before I met him, which made it easier to follow his story.

While no one else objected to the tape recorder as such, some had made me shut it off at certain points during the interview. Interestingly, all of them were rescuers. One of them, Janina Morawska, a Yad Vashem recipient, took into her one-room hut a Jewish woman, a stranger whom she kept for 9 months. Asked whether this woman was grateful, Morawska first answered in the affirmative, but then made me stop taping. Only then with a sad and bitter smile she added: "Yes, she was grateful, mostly with words. Twice, and only twice, did she send me $10." Hela Horska, also a Yad Vashem recipient, protected fourteen Jews. Hers was an unusual case because these Jews were very religious. Her husband was a doctor and she knew them only as his patients. When asked about the gratitude shown by these people, she too made me stop the recorder and said: "After all, they were inferior to us. They did not know any better. From people like this I expected nothing. So why should I be disappointed?" Without exception, when criticizing the rescuers made me stop taping. Such criticisms however were rare, with the rescuers showing a greater willingness to criticize their government than their Jewish charges. To preserve all materials, after each interview as soon as I was alone, I wrote down my personal observations and comments.

Each interview lasted from two to eight hours. In a few instances I met with a person more than once. I tried to set up another meeting when I felt that the individual was withholding some information that might be useful. For example this happened with Wacka Nowak. She had been protected by a Polish professor who had treated her with much consideration. During the interview, however, Wacka became more rather than less nervous. She stopped and took a drink, yet she remained tense. When I asked why she had not maintained contact with her benefactors, she became very agitated. Finally, she said: "That is all I am prepared to tell you. Better ask different questions!" I knew enough not to insist, yet I also knew that she was withholding something. Before I left, I asked for a follow-up interview. In our second meeting I learned the reason for her reaction. Wacka, fifteen at the time and beautiful, was in Warsaw with the professor and his mother during the Polish uprising; the professor's wife was in the country with the children. The professor, his mother, and Wacka were evacuated from Warsaw with the rest of the population. They were moved to a primitive village, where all three had to share one bed. During the night the professor, an "old man" of forty-five or so, made advances to Wacka and became furious when she rejected him. After a few days he and his mother left the young girl behind, all alone among strangers. It was hard enough to feel unjustly abandoned, but Wacka also soon learned that before his departure the professor had told the peasant in whose hut they lived that Wacka was Jewish and that she had joined them by chance. Fortunately, the girl succeeded in convincing the peasant that this was a lie.

Because of the very personal nature of the experiences I was inquiring about, I attempted to conduct each interview in private. As much as possible I wanted to avoid any inhibition of, or outside influence on, a person's reactions. I was not always successful. The Polish rescuer, Eva Anielska, arranged an interview for me with Marek Dunski, a prominent Catholic writer, who was responsible for rescuing hundreds of Jewish children. Even though my previous meeting with Anielska had been private, the second interview she set up as a semisocial event. Besides myself, the hostess, and Dunski, there were two more Polish friends, both of whom had some connection to the rescue of Jews. On arriving, I expressed my interest in interviewing the guest writer and took out my schedule, pencil, and tape recorder, but it was of no use. While everyone talked, I juggled questions and answers with Dunski and listened to the others as well. All this caused the meeting to last much longer, but in the end I seem to have gotten the information I wanted.

Once in a while in the course of an interview a member of the family would join in. At such times I would stop the questions and resume them after the intruder had left. I remember one unusual meeting in the elegant study of a prosperous businessman who was a Jewish survivor. After half an hour his wife returned home. As soon as she entered the room, I stopped recording. After we were introduced, she sat down, lit a cigarette, and began to chat. I realized that she had no intention of leaving but was scrutinizing me as if waiting for me to start my questions. After a brief but embarrassing silence, I explained, as gently as possible, that I had to conduct the interview in private. Visibly annoyed, she left in a huff, only to return in half an hour to ask how much longer the interview would take.

Another thirty minutes passed and this time the wife returned complaining that she should have been warned about the length of the interview. She had turned down an invitation to a friend's house in order to spend the evening with her husband. Without waiting for a reaction she left abruptly. But this was not the end. During her next appearance she announced that a great movie was being shown on television. We had to move to the kitchen because the room we were in had the TV set.

Throughout these interruptions her husband said nothing. Each time the wife left he continued the story as if nothing had happened. He seemed totally absorbed in what he was telling me. As far as I could tell, his wife's appearances and comments left no impression. This was a long and fruitful interview. As I was leaving he thanked me, telling me how grateful he was for the opportunity to talk.

With interviews more so than with the other sources, one might legitimately wonder how much credence can be given to accounts told more than thirty years after the events? How much does one remember? Still, such events as these were the most dramatic and painful of a person's life, and as such are not apt to be forgotten. Time I believe is most likely to cloud our memory for common events but less likely to rob us of extraordinary experiences.

Those of us who survived the Holocaust have lived with our memories and know that they cannot be erased. I could therefore argue, with some justification, that these experiences are firmly entrenched in our memory. Besides, given the shortcomings of the available material I had to choose between no information at all on vital issues, or information that might be tarnished by time. I decided in favor of information. In some instances I was able to compare an interview to other earlier and independent sources. Thus, without exception for all the Christians I interviewed, I had also access to their testimonies in archives. For some I had a third independent source, in a person's published memoirs. Most of their accounts reached as far back as 1945–46; some are more recent. For Holocaust survivors, except for two, none had ever testified before, and only a few had written their memoirs.

Also, in quite a number of cases, I had information about a single individual from three or more separate sources: testimony, published memoirs, and direct interview. Again, this was especially the case with Poles and less frequently with Holocaust survivors.

Thus, for example, during the war Paweł Remba was a member of the Security Corps, a part of the Polish underground. In this capacity he was involved in smuggling Jews out of the Warsaw ghetto while the uprising was in progress. During one such action he was injured and to this day he walks with a limp. Having first read about him in a book, I interviewed him in 1979. Although his testimony was dated 1957, the same detailed information reappeared in the interview. Indeed, except for the additional specific information I elicited for my study, there was very little difference between these sources.

I also had the opportunity to interview six pairs: the rescued and their rescuers. Except for one pair, the rest each lived in different countries, either Poland and the United States or Poland and Israel. As a rule, when comparing information about such pairs I found omissions but no serious distortions. When distortions did occur, they involved the chronology of events, special dates, or duration of events, and as such were not significant for my purposes. As a rule, most interviews suggest a high level of accuracy.

With thirty-four Jews and thirty-one Poles I stopped my interviewing. As I had expected, by far the richest and most extensive evidence came from these direct interviews. Altogether I collected data from a variety of sources: published accounts and memoirs, unpublished testimonies, and direct in-depth interviews.[11] From these varied sources I collected direct evidence from 189 Poles and 308 Jews. These Jewish survivors offered adequate descriptions about 565 Polish helpers.[12]

Eventually all cases were codified and quantified.[13] Both unevenness and variability of data required that some of it, the interviews in particular, be used for qualitative analysis as well.

I hope that by mixing numbers with descriptions of detailed life experiences I have reconciled and achieved a certain balance between an observer and a participant, between objectivity and involvement.

Notes

Introduction

1. "Frank, Anne," *Encyclopedia Judaica*, Vol. 7 (Jerusalem: Keter Publishing House, Ltd., 1973), pp. 53–54.

2. Philip Friedman, *Their Brothers' Keepers* (New York: Holocaust Library, 1978), p. 63.

3. "Readers' Supplement," in *Anne Frank: The Diary of a Young Girl* (New York: Washington Square Press, 1972), pp. 1–53. Names of these protectors were: Mr. Koophuis, Mr. Kraler, Miep van Santen, and Elli Vossen (p. 14). According to the official Yad Vashem records, the names of these five righteous Christians were: Jan and Miep Gies, Jo Kleiman, Victor Kugler, and Elisabeth Van-Wijk-Voskuyl. Such name differences might be caused by transformations from Dutch into Hebrew and English.

4. Personal communication, Dr. Mordechai Paldiel, director of the Department for the Righteous, Yad Vashem, Jerusalem, Israel.

5. Ibid.

6. See Peter Hellman, *Avenue of the Righteous* (New York: Atheneum, 1980), p. ix. Also, the *Encyclopedia Judaica* mentions four kinds of distinctions, by differentiating between a "certificate of honor" and "a letter of esteem." See *Encyclopedia Judaica*, Vol. 14, p. 184. The kinds of distinctions are by no means set. Discussions about their number and quality are still continuing. Personal communication, Dr. M. Paldiel, director of the Department for the Righteous, Yad Vashem, Jerusalem, Israel.

7. Personal communication, Dr. Paldiel.

8. Moshe Bejski, "The Righteous among the Nations and Their Part in the Rescue of Jews," in Yisrael Gutman and Livia Rothkirchen (eds.), *The Catastrophe of European Jews* (Jerusalem: Yad Vashem, 1976), pp. 582–607.

9. Władysław Bartoszewski and Zofia Lewinówna, *Ten Jest Z Ojczyzny Mojej* [This One Is from My Country] (Kraków: Wydawnictwo Znak, 1969); Arieh Bauminger, "Righteous Gentiles," *Jewish Spectator* 27–31 (September 1964): 9–10; P. Friedman, *Their Brothers' Keepers;* Kurt R. Grossman, "The Humanitarian That Cheated Hitler," *Coronet*, September 1959, pp. 66–71; Michael Horbach, *Out of the Night* (New York: Frederick Fell Inc., 1967); H.D. Leuner, *When Compassion*

Was a Crime (London: Oswald Wolf, 1966); Felix Kersten and Herman Briffault, *The Memoirs of Dr. Felix Kersten* (New York: Doubleday, 1947); Władysław Smolski, *Zaklęte Lata* [Cursed Years] (Warszawa: Pax, 1964).

10. A few such examples are: Corrie ten Boom, with John and Elizabeth Sherrill, *The Hiding Place* (Old Tappan, N.J.: Fleming H. Revell Co., 1971); Phillip Hallie, *Lest Innocent Blood Be Shed* (New York: Harper and Row, 1979); Hellman, *Avenue of the Righteous;* Kazimierz Iranek-Osmecki, *He Who Saves One Life* (New York, Crown: Publishers, Inc., 1971); Alexander Ramati, *The Assisi Underground: The Priests Who Rescued Jews* (New York: Stein and Day Publishers, 1978).

11. Per Anger, *With Raoul Wallenberg in Budapest* (New York: Holocaust Library, 1981); John Bierman, *Righteous Gentile: The Story of Raoul Wallenberg, Missing Hero of the Holocaust* (New York: The Viking Press, 1981).

12. Thomas Keneally, *Schindler's List* (New York: Simon and Schuster, 1982).

13. Copies of these different laws are included in Raul Hilberg, ed., *Documents of Destruction* (New York: Quadrangle Books, 1971).

14. Raul Hilberg, *The Destruction of the European Jews* (New York: New Viewpoints, 1973), pp. 257–267; Filip Friedman, "Zagłada Żydów Polskich W Latach 1939–1945" [Destruction of the Polish Jewry], *Biuletyn Głównej Komisji Badania Zbrodni Niemieckiej W Polsce* no. 6 (1946): 165–208.

15. Lucy S. Dawidowicz, *The War against the Jews 1939–1945* (New York: Holt, Rinehart and Winston, 1975), p. 372.

16. Hilberg, *Destruction of the European Jews*, p. 357; Hugo Valentin, "Rescue and Relief Activities in Behalf of Jewish Victims of Nazism in Scandinavia," *YIVO Annual of Social Science* 8 (1953): 224–251.

17. Henry L. Mason, "Jews in the Occupied Netherlands," *Political Science Quarterly* 99, no. 2 (Summer 1984): 339.

18. Harold Flender, *Rescue in Denmark* (New York: Simon and Schuster, 1963), p. 259; Leni Yahil, *The Rescue of Danish Jewry, Test of Democracy* (Philadelphia: The Jewish Publication Society of America, 1969), pp. 226–227.

19. Hilberg, *Destruction of the European Jews*, p. 363.

20. This, of course, does not mean that Nazi collaborators were completely absent. See Helen Fein, *Accounting for Genocide* (New York: The Free Press, 1979), pp. 144–152.

21. Nora Levin, *The Holocaust* (New York: Schocken Books, 1973), p. 404.

22. B.A. Sijes, "Several Observations Concerning the Position of the Jews in Occupied Holland during World War II," pp. 527–553 in *Rescue Attempts during the Holocaust*, eds. Yisrael Gutman and Efraim Zuroff (Jerusalem: Yad Vashem, 1977), p. 548.

23. Mason, "Jews in the Occupied Netherlands," p. 330.

24. Gideon Hausner, *Justice in Jerusalem* (New York: Holocaust Library, 1966), pp. 259–260.

25. Fein, *Accounting for Genocide*, p. 287; Jacob Presser, *The Destruction of the Dutch Jews* (New York: E.P. Dutton and Co., Inc., 1969), p. 327.

26. For examples of this aid see Haim Auni, "The Zionist Underground in Holland and France and the Escape to Spain," pp. 555–590 in Gutman and Zuroff, eds., *Rescue Attempts; Marga Minco, Bitter Herbs* (New York: Oxford University Press, 1960); Leesha Rose, *The Tulips Are Red* (New York: A.S. Barnes and Co., 1978).

27. Fein, *Accounting for Genocide*, p. 73.

28. Minco, *Bitter Herbs*, pp. 8–9; Presser, *Destruction of the Dutch Jews*, pp. 325–327.

29. Hilberg, *Destruction of the European Jews*, p. 378.

30. Sijes, "Several Observations," p. 528. According to Filip Friedman, 40,000 Dutch Jews tried to survive by passing and 15,000 succeeded in doing so. See Friedman, *Their Brothers' Keepers*, p. 63.

31. Dawidowicz, *War against the Jews*, p. 402; Sijes, "Several Observations," p. 553. Unlike Dawidowicz, Sijes maintains that not 75 percent but over 80 percent of the Jews in Holland perished.

32. Hilberg, *Destruction of the European Jews*, p. 433.

33. Dawidowicz, *War against the Jews*, p. 387.

34. Helen Fein ranks the success of the anti-Semitic movement in Bulgaria as moderate. See Fein, *Accounting for Genocide*, p. 72.

35. Hilberg, *Destruction of the European Jews*, p. 474.

36. Ibid., p. 481.

37. Friedman, *Their Brothers' Keepers*, p. 105.

38. Fein, *Accounting for Genocide*, p. 115.

39. Hilberg, *Destruction of the European Jews*, pp. 481–483.

40. Dawidowicz, *War against the Jews*, pp. 387–389; Hilberg, *Destruction of the European Jews*, pp. 481–483.

41. Levin, *Holocaust*, p. 549.

42. See Friedman, "Destruction of the Polish Jewry."

43. My research data and methodology are described in the Postscript, preceding these notes.

44. Taking an overview of righteous Christians, Professor Baron asserts that "scholarly study of the 'Righteous Gentiles' has been one of the least adequately researched aspects of the Holocaust." See Lawrence Baron, "The Holocaust and Human Decency: Profiling the 'Righteous Gentiles'," unpublished paper (1983).

45. Jacob Lestchinsky, "The Jews in the City of the Republic of Poland," *YIVO Annual of Jewish Social Science* 1, (1946): 156–177.

46. Jacob Lestchiñsky, "The Industrial and Social Structure of the Jewish Population of Interbellum Poland," *YIVO Annual of Jewish Social Science* 2 (1956–1957): 246.

47. L. Lifschutz, "Selected Documents Pertaining to Jewish Life in Poland, 1919–1938," in *Studies on Polish Jewry*, ed. Joshua A. Fishman (New York: YIVO Institute for Jewish Research, 1974), p. 280.

48. Antony Polonsky, *Politics in Independent Poland, 1921–1939* (Oxford: Clarendon Press, 1972), p. 40.

49. Celia S. Heller, *On the Edge of Destruction* (New York: Columbia University Press, 1977), p. 69.

50. Ibid., p. 17.

51. Ibid., pp. 17–21.

52. Edward D. Wynot, Jr., *Warsaw between the World Wars: Profile of the Capital City in a Developing Land* (New York: Columbia University Press, 1983), pp. 1–10.

53. Heller, *Edge of Destruction*, p. 50.

54. Lifschutz, "Selected Documents," p. 277.

55. Up to World War II Poland's Communist party was small, numbering not more than 20,000 members. In this Communist party Jews held most of the leader-

ship positions. Considering, however, that there were 3.5 million Jews and the small number of Communists, the accusation that Jews were Communists is hardly justified. For a discussion of Communists in prewar Poland see Alicja Iwanska, *Contemporary Poland, Society, Politics, Economy* (Chicago: The Chicago University Press, 1955), p. 288; Paul Lendvai, *Anti-Semitism without Jews* (New York: Doubleday, 1971), pp. 56–63.

56. These rights were reiterated in Poland's 1921 Constitution. But neither the provisions of the Versailles Minority Treaty nor Article 109 of the 1921 Constitution supporting these rights and freedoms became a reality. For a discussion of these rights see Jacob Lestchiński, "Economic Aspects of Jewish Community Organization in Independent Poland," *Jewish Social Studies* 9, no. 1–4 (1947): 319; Lifschutz, "Selected Documents," p. 277; Bernard D. Weinryb, "The Jews in Poland," in *The Jews in the Soviet Satellite*, eds. Peter Meyer, et al. (Westport, Conn.: Greenwood Press, 1953), p. 208; Edward D. Wynot, Jr., *Polish Politics in Transition* (Athens, Ga.: University of Georgia Press, 1974), p. 18.

57. Heller, *Edge of Destruction*, p. 57.

58. Polonsky, *Politics in Independent Poland*, p. 71.

59. Wynot, *Polish Politics*, p. 48.

60. Edward D. Wynot, Jr., "A Necessary Cruelty: The Emergence of Official Anti-Semitism in Poland, 1936–1939," *The American Historical Review* 76, no. 4 (October 1971): 1035–1058.

61. The rightists consisted of many different political shadings. All of them were anti-Semitic, but the strength of their anti-Semitism varied. Also, neither the socialist nor the peasant parties were homogeneous. Each varied in the extent to which it adhered to leftist principles and in its position toward anti-Semitism.

62. Lendvai, *Anti-Semitism without Jews*, p. 204.

63. Wynot, "Necessary Cruelty," p. 1035.

64. Because of the Versailles Minority Treaty and the 1921 Constitution guaranteeing the rights and freedoms of the minorities, these measures never became laws. This, however, by no means diminished their discriminatory effectiveness.

65. Lestchiński, "Structure of Jewish Population," p. 245; Emmanuel Ringelblum, *Polish–Jewish Relations during the Second World War* (Jerusalem: Yad Vashem, 1974), pp. 10–22.

66. Lifschutz, "Selected Documents," p. 279.

67. During this pogrom two Jews and one Pole died. See Paweł Korzec, "Anti-Semitism in Poland as an Intellectual, Social and Political Movement," in Fishman, ed., *Studies on Polish Jewry*, p. 87; Lifschutz, "Selected Documents," p. 279.

68. These two posters are a part of the YIVO Territorial Collection on Poland. See YIVO Territorial Collection on Poland, Vilno Archives Record Group, no. 28 (1931), YIVO Archives, New York City. From the second poster I took only the picture and one caption, leaving the rest out. The text also urges the Poles to support the business boycott.

69. Lifschutz, "Selected Documents," p. 280.

70. Jacob Lestchiński, *Crisis, Catastrophe and Survival: A Jewish Balance Sheet, 1914–1948* (New York: Institute of Jewish Affairs of the World Jewish Congress, 1948), p. 33.

71. Carole Fink, "Germany and the Polish Elections of November 1930: A Study in League Diplomacy," *East European Quarterly* 15, no. 2 (June 1981): 181–207; Wynot, *Polish Politics*, p. 21.

72. Korzec, "Anti-Semitism in Poland," p. 91.

73. For a historical account of this plan see Philip Friedman, "The Lublin Reservation and the Madagascar Policy," *YIVO Annual of Social Science* 8 (1958): 151–177.

74. Weinryb, "Jews in Poland," p. 210.

75. Bernard Wasserstein, *Britain and the Jews of Europe, 1939–1945* (London: Institute for Jewish Affairs, 1979), p. 19.

76. Lestchińsky, *Crisis, Catastrophe and Survival*, pp. 20–21.

77. Bernard Goldstein, *The Stars Bear Witness* (New York: The Viking Press, 1949), p. 52; Chaim Kapłan, *Scrolls of Agony, The Warsaw Diary of Chaim Kapłan* (New York: The Macmillan Co., 1956), p. 51; Raul Hilberg, et al., eds., *The Diary of Adam Czerniakow* (New York: Stein and Day Publishers, 1979), pp. 131–133; Emmanuel Ringelblum, *Polish-Jewish Relations in the Second World War*, edited by Jósef Kermisz and Shmuel Krakowski (Jerusalem: Yad Vashem, 1974), pp. 51–53.

78. For the Nazis the definition of a Jew created many difficulties. For years they continually changed the definition. Needless to say, this applied particularly to part Jews, the *Mischlinge*. For a discussion of these issues see Hilberg, *Destruction of the European Jews*, pp. 44–53.

79. See Dawidowicz, *War against the Jews*, pp. 223–260; Philip Friedman, "Aspects of Research on the Judenrat," pp. 539–553 in *Roads to Extinction: Essays on the Holocaust*, ed. June Ada Friedman (Philadelphia: The Jewish Publications Society of America, 1980); Hilberg, *Destruction of the European Jews*. Hilberg discusses the Jewish Councils in many parts of the book. His attitudes toward these councils are by and large unsympathetic. For a more balanced view see Isaiah Trunk, *Judenrat* (New York: Macmillan, 1972.)

80. These are only approximate figures. For a thorough discussion of the difficulties in arriving at the exact number of inhabitants when the Lodz ghetto was formed see Lucjan Dobroszycki, ed., *The Chronicle of the Lódź Ghetto, 1941–1944* (New Haven, Conn.: Yale University Press, 1984), pp. xxxvii–xxxix.

81. Philip Friedman, "Mordechai Chaim Rumkowski of Lódź," pp. 353–364 in J.A. Friedman, ed., *Roads to Extinction*.

82. Emmanuel Ringelblum, *Notes from the Warsaw Ghetto* (New York: Schocken Books, 1975).

83. Stanisław Adler, *In the Warsaw Ghetto* (Jerusalem: Yad Vashem, 1982), p. 323.

84. Yitzhak Katzenelson, *Vittel Diary* (Israel: Ghetto Fighters' House, 1972), p. 63.

85. Yitskhok Rudashevsky, *The Diary of the Vilna Ghetto, June 1941–April 1943* (Israel: Ghetto Fighters' House, 1973), pp. 32–33.

86. Katzenelson, *Vittel Diary*, pp. 224–225.

87. Ringelblum, *Notes from Ghetto*, p. 205.

88. From now on throughout the book, quotes not identified by specific sources refer to the rescuers and the rescued whom I interviewed specifically for this study. For a description of the research and methodological aspects of this study see the Postscript.

89. Adolf Berman, "The Fate of the Children in the Warsaw Ghetto," pp. 400–421 in Gutman and Rothkirchen, eds., *Catastrophe of European Jewry;* Lucy S. Dawidowicz, ed., *A Holocaust Reader* (New York: Behrman House, Inc., 1976), pp. 193–194.

90. For documents listing welfare activities see Dawidowicz, ed., *A Holocaust Reader*, pp. 179–193; see also Mark Dworzecki, "The Day to Day Stand of the

Jews," pp. 367–399 in Gutman and Rothkirchen, eds., *Catastrophe of European Jewry;* Ludwig Hirszfeld, *Historia Jednego Życia* [The story of one life] (Warszawa: Pax, 1957). Before he left the ghetto Hirszfeld, a famous doctor and converted Jew, tried courageously to contain the Warsaw ghetto epidemics.

91. Dawidowicz, *Holocaust Reader,* p. 207; Martin Gray and Max Gallo, *For Those I Loved* (Boston: Little, Brown and Co., 1972).

92. Hilberg, *Destruction of the European Jews,* pp. 308–345. A set of documents about death by starvation were published right after the war, based on the actual experience of the ghetto inhabitants. See *Choroba Głodowa* [Starvation illness] (Warsaw: American Joint Distribution Committee, 1946).

93. Hilberg, *Destruction of the European Jews,* pp. 309–332; Alexander Donat, ed., *The Death Camp Treblinka* (New York: Holocaust Library, 1979); Miriam Novitch, *Sobibor* (New York: Holocaust Library, 1980).

94. Yitzhak Arad, et al., eds., *Documents of the Holocaust* (Jerusalem: Yad Vashem, 1981), pp. 279–282; Raul Hilberg, et al., *The Warsaw Diary of Adam Czerniakow* (New York: Stein and Day Publishers, 1979).

95. Friedman, "Destruction of the Polish Jewry."

96. Dawidowicz, ed., *Holocaust Reader,* p. 67.

97. Władysław Bartoszewski, "Egzekucje Publiczne W Warszawie W Latach, 1943–1945" [Public executions in Warsaw], *Biuletyn Głównej Komisji Badania Zbrodni Niemieckiej W Polsce,* no. 6 (1946): 211–224; Ringelblum, *Notes from Ghetto,* p. 236; Tatiana Berenstein, et al., eds., *Exterminacja Żydow Na Ziemiach Polskich W Okresie Okupacji Hitlerowskiej* [Jewish extermination in Poland during Hitler's occupation] (Warszawa: Żydowski Instytut Historyczny, 1957), pp. 121–122.

Chapter 1

1. Michael Borwicz has described life on the Aryan side in three volumes of *Arische Papirn* [Aryan papers], in Yiddish in Buenos Aires. In addition to these three volumes published by the Association of Polish Jews in Argentina (1955), Borwicz also wrote a book in French on the same topic, *Les Vies Interdites* [Forbidden lives] (Paris: Casterman, 1969). These books describe what life was like on the Aryan side.

2. Tatiana Berenstein and Adam Rutkowski, *Assistance to the Jews in Poland* (Warsaw: Polonia Foreign Languages Publishing House, 1963).

3. Lucy Dawidowicz, ed., *A Holocaust Reader* (New York: Behrman House, Inc., 1976), p. 67; Tatiana Berenstein, et al, eds., *Exterminacja Żydow Na Ziemiach Polskich Zbiór Dokumentów* [Extermination of Jews on Polish land during the German occupation], collection of documents (Warszawa: Żydowski Instytut Histozyczny, 1957), p. 123.

4. Emmanuel Ringelblum, *Notes from the Warsaw Ghetto* (New York: Schocken Books, 1975), pp. 236–237.

5. Władysław Bartoszewski, "Publiczne Egzekucje W Warszawie W Latach 1943–1944," [Public executions in Warsaw], *Biuletyn Głownej Komisji Badania Zbrodni Niemieckiej W Polsce,* no. 6 (1946): 211–224; Philip Friedman, *Their Brothers' Keepers* (New York: Holocaust Library, 1978); Kazimierz Iranek-Osmecki, *He Who Saves One Life* (New York: Crown Publishers, Inc., 1971); Szymon Datner, *Las Sprawiedliwych* [The forest of the righteous] (Warszawa: Książka I Wiedza, 1968).

6. Depending on the sources there are slight variations in the estimated number of Jewish survivors in Poland. For comparisons of these figures as well as

comparisons with the destruction of other European Jews, see Lucy S. Dawidow-icz, *The War against the Jews, 1933–1945* (New York: Holt, Rinehart and Winston, 1975), p. 403; Raul Hilberg, *The Destruction of the European Jews* (New York: New Viewpoints, 1973), p. 767; Jacob Lestchińsky, "Balance Sheet of Extermina-tion," *Jewish Affairs* 1, no. 1 (February 1946): 10. In contrast to the preceding, Friedman focuses on Polish Jews only. See Filip Friedman, "Zagłada Żydów Pol-skich W Latach 1939–1945" [Destruction of the Polish Jewry], *Biuletyn Głównej Komisji Badania Zbrodni Niemieckiej W Polsce*, no. 6 (1946): 165–203.

7. Personal communication by the Polish–Jewish historian Szymon Datner dur-ing a seminar at YIVO (Institute of Jewish Research, New York City), 1980.

8. Bernard D. Weinryb, "The Jews in Poland," pp. 207–326 in *The Jews in the Soviet Satellite*, eds. Peter Meyer, et al. (Westport, Conn.: Greenwood Press, 1953).

9. This study and the many issues related to the research methods are de-scribed in the Postscript.

10. This story was related by Dora Frank, a survivor I interviewed in New York. During the war she passed as Christian and worked as a governess on a country estate where Ignac worked as an assistant to the forester. Dora learned about Ignac's true identity only at his death.

11. Of the 308 survivors included in my study, 48 percent had papers that identified them as Christians and 52 percent did not.

12. The following tabulation shows percentages of survivors hiding or passing who had false documents.

Status	With documents	Without documents
Hiding among Christians	23%	85%
Passing as Christians	77%	15%
Total*	$n = 146^\dagger$	$n = 124$

*The total does not add up to 308 because for some this kind of information is missing.
†The *n* always refers to the number of cases.

13. Władysław Bartoszewski and Zofia Lewinówna, *Ten Jest Z Ojczyzny Mojej* [This one is from my country] (Kraków: Wydawnietwo Znak, 1969), pp. 472–474.

14. Among the survivors in my study, 54 percent are women and 46 percent men. Of the women, 79 percent looked Polish compared to 56 percent of the men. Similarly, whereas among the women only 10 percent used outright faulty Polish, the same was true of 29 percent of men. See Nechama Tec, "Sex Distinctions and Passing as Christians during the Holocaust," *East European Quarterly* 18, no. 1 (March 1984): 113–123.

15. Sixty percent of the men were hidden, compared to 47 percent of women. Also when compared in terms of employment, 48 percent of Jewish women worked and only 26 percent of Jewish men who lived on the Aryan side. See ibid.

16. Felix Kanabus, "Address of the JNF September 20, 1964," pp. 392–395 in *Anthology of Holocaust Literature*, eds. Jacob Gladstein, et al., (Philadelphia: The Jewish Publication Society of America, 1969).

17. Lucjan Dobroszycki describes an underground group of Polish doctors who during the occupation extended medical help to Jews. Among the doctors he identi-fied Estowski by his real name; see pp. 241–244 in Bartoszewski and Lewinówna, *This one is from my country*.

18. Balicka Helena Kozłowska, *Mur Miał Dwie Strony* [The wall had two sides] (Warszawa: Wydawnictwo Ministerstwa Obrony Narodowej, 1958), pp. 13–14.

19. Celia S. Heller, "Poles of Jewish Background—The Case of Assimilation without Integration in Prewar Poland," pp. 242–276 in *Studies on Polish Jewry*, ed. Joshua Fishman (New York: YIVO Institute for Jewish Research, 1974).

20. Antony Polonsky, *Politics in Independent Poland, 1921–1939* (Oxford: The Clarendon Press, 1972), p. 40.

21. Of the 308 survivors in my study, 14 percent spoke a faulty Polish while 24 percent seemed to speak the language fluently. The majority seemed to know Polish fairly well but could be recognized by their speech.

22. Of the survivors I studied, 20 percent conformed to the Semitic look. As for religiosity, 56 percent may be classified as not religious, and 8 percent as very religious.

Chapter 2

1. This letter is included in Tatiana Berenstein, et al., eds., *Exterminacja Żydów Na Ziemiach Polskich Zbiór Dokumentów* [Extermination of Jews on Polish land during the German occupation], collection of documents (Warszawa: Żydowski Instytut Historyczny, 1957), p. 529.

2. For documents listing such rewards see Berenstein, et al., *Extermination of Jews*. For comments about these, see Zygmunt Klukowski, *Dziennik Z Lat Oku-pacji, Zamojszcyzny* [Wartime diary from Zamość] (Lublin: Lubelska Spółdzielnia Wydawnicza, 1958); Stanisław Zieminski, "Kartki Dziennika Nauczyciela W Lu-kowie Z Okresu Okupacji Hitlerowskiej" [Notes from a teacher's diary from Lukow during Hitler's occupation], *Biuletyn Żydowskiego Instytutu Historycznego*, no. 27 (1958): 105–112.

3. Ludwik Hirszfeld, *Historia Jednego Życia* [The story of one life] (Warsaw: Pax, 1957), p. 407.

4. Klukowski, *Wartime Diary*, p. 299.

5. Ziemiński, "Notes from a Teacher's Diary," p. 109.

6. Nathan Gross, "Days and Nights in the 'Aryan' Quarter: The Daily Worries of a Jew Carrying Aryan Papers," *Yad Vashem Bulletin*, no. 4/5 (1959): 12–13; Kazimierz Iranek-Osmecki, *He Who Saves One Life* (New York: Crown Publishers, Inc., 1971), pp. 52–53; "Polish Underground Liquidates Betrayers of Jews," *Ghetto Speaks*, no. 27 (July 15, 1944); Emmanuel Ringelblum, *Polish–Jewish Relations during the Second World War* (Jerusalem: Yad Vashem, 1974), pp. 100–139. These are just a few of the many publications that discuss Polish collaborators.

7. W. Zagórski, *Wolność W Niewoli* [Freedom in slavery] (London: Nakładem Autora I Czytelników Przedstawiciela, 1971), p. 479.

Chapter 3

1. Lucy S. Dawidowicz, *The War against the Jews, 1933–1945* (New York: Holt, Rinehart and Winston, 1975), p. 395.

2. Here are a few of the many sources that describe what life was like for the Poles under the Nazi occupation: Władysław Bartoszewski, *1859 Dni Warszawy* [1859 Warsaw days] (Kraków: Wydawnictwo Znak, 1974); Władysław Bartoszewski, *Straceni Na Ulicach Miasta* [Perished in the streets of the city] (Warszawa: Książka I Wiedza, 1970); Ludwik Landau, *Kronika Lat Wojny I Okupacji* [War chronicle], 3 vols. (Warszawa: Państwowe Wydawnictwo Naukowe, 1962); Lańdau perished while trying to survive by passing. Stefan Korboński, *The Polish Underground*

State: A Guide to the Underground (Boulder, Colo.: East European Quarterly, 1978); Stanisław Wroński and Maria Zwolakowa, *Polacy I Żydzi, 1939–1945* [Poles and Jews] (Warszawa: Książka I Wiedza, 1971).

3. Nora Levin, *The Holocaust* (New York: Schocken Books, 1973), p. 163.

4. Władysław Bartoszewski, *The Blood Shed Unites Us* (Warsaw: Interpress Publishers, 1970), p. 229.

5. Levin, *Holocaust*, p. 163; Wroński and Zwolakowa, *Poles and Jews*, p. 450.

6. Stefa thought that initially her husband might have been paid by Laminski for accepting his mistress. Money she felt was her husband's reason for accepting Irena.

7. For an excellent account and study of the Polish Warsaw uprising, see Janusz Z. Zawodny, *Nothing but Honor: The Story of the Warsaw Uprising, 1944* (Stanford, Calif.: Hoover Institution Press, 1979).

8. After coming across Irena's testimony, I asked Stefa Dworek's permission to use it. By using it I am revealing the real name of Mrs. Dworek which is Stanisława Davidżiuk. She answered me promptly, allowing me to use her name whenever I choose. Even though the real name of this rescuer and the rescued can be easily traced through the reference below, in the text in line with my established procedure I prefer to use a fictitious name.

9. Władysław Bartoszewski and Zofia Lewinówna, *Ten Jest Z Ojczyzny Mojej* [This one is from my country] (Kraków: Wydawnictwo Znak, 1969), pp. 582–584.

10. Ludwik Hirszfeld, *Historia Jednego Życia* [The story of one life] (Warszawa: Pax, 1957), p. 424.

11. Izabella Stachowicz Czajka, *Ocalił Mnie Kowal* [I was saved by a blacksmith] (Warszawa: Czytelnik, 1956).

12. Oskar Pinkus, *The House of Ashes* (New York: The World Publishing House Co., 1964), p. 198.

13. Ibid., p. 223.

14. I am reiterating what I have already said in the introductory chapter. For a comprehensive treatment of this issue, see Helen Fein, *Accounting for Genocide* (New York: The Free Press, 1979).

15. Lucy S. Dawidowicz, ed., *A Holocaust Reader* (New York: Behrman House Inc., 1976), p. 67.

16. In some Polish sources it has been asserted that Poland was the only European country where an official law demanded the death sentence for saving Jews. See Bartoszewski, *Blood Shed Unites Us*, p. 227; Tatiana Berenstein and Adam Rutkowski, "O Ratowaniu Żydow Przez Polaków W Okresie Okupacji Hitlerowskiej" [Saving Jews by Poles during Hitler's occupation], *Biuletyn Żydowskiego Instytutu Historycznego*, no. 35 (1960): 3–46; Wroński and Zwolakowa, *Poles and Jews*, p. 402. The document in the text was excerpted from Wroński and Zwolakowa, *Poles and Jews*, p. 436. Note that not all executions had to do with the rescuing of Jews.

17. In the protectorate of Bohemia and Morawia, in the wake of the Heydrich assassination in 1942, Kurt Daluege, the new reich protector, "issued an ordinance establishing the death penalty for anyone aiding or failing to report persons engaged in activities hostile to the Reich, including sheltering Jews." See Livia Rothkirchen, "Czech Attitudes towards the Jews during Nazi Regime," *Yad Vashem Studies* 13 (1979): 314. This information is especially important, since the Czechs generally had a positive record of Jewish toleration. Still, 90 percent of the Czech Jews died and, according to Rothkirchen, only 424 survived in hiding (p. 315). I am grateful to Lawrence Baron for bringing these facts to my attention.

I also found a reference to Norway stating that "the Germans made public announcements in Norwegian that anyone extending aid to Jews in the way either of clothing, food or shelter would be liable to execution together with his family." Perhaps the phrase "would be liable" suggests that execution was only a possibility, because the statement continues: ". . . several hundred patriots were interned in the Grini concentration camp." Arieh L. Bauminger, *Roll of Honor* (Jerusalem: Yad Vashem, 1970), p. 64.

18. Wroński and Zwolakowa, *Poles and Jews,* p. 417.

19. I have translated freely from the Polish. See Bartoszewski and Lewinówna, *This One Is from My Country,* pp. 855–859.

20. I have translated freely from the Polish. Ibid., pp. 862–865.

21. Ibid., p. 868.

22. Philip Friedman, *Their Brothers' Keepers* (New York: Holocaust Library, 1978), p. 127.

23. Wroński and Zwolakowa, *Poles and Jews,* p. 435.

24. Ibid., p. 433.

25. Of the 189 rescuers in my sample, 60 percent were married, 7 percent widowed, and 33 percent single. Also, 43 percent of them had children and 57 percent were childless.

26. Considered in terms of family support for rescue, the information is as follows:

Family situation	Percent
Family supports efforts to save Jews	60
Family opposes, disapproves of efforts to save Jews	12
Live alone	28
	$n = 144$

27. Staszek's case is described in Ruth M. Gruber, "The Pole Who Saved 32 Jews," *Hadassah Magazine,* December 1968, pp. 21, 36.

Chapter 4

1. Of the survivors, 62 percent say that they were the ones who asked for help. In 20 percent of the cases they were directed by someone else to the helper. Only in 15 percent of the cases did the Poles themselves initiate their aid.

2. Of the survivors, 76 percent say that the help they received was not promised to them ahead of time. To be sure, about 50 percent of the survivors were refused help at one time or another. It is important to emphasize that I am dealing here only with survivors; no doubt many of those who tried to survive by passing perished because they failed to find Poles who would help them.

3. Of these Poles 29 percent initiated their help to Jews, while 30 percent say that the Jews approached them for help. Help was initiated through intermediaries in 28 percent of cases, and in 13 percent information about the way the help started is unavailable.

4. Szymon Datner, *Las Sprawiedliwych* [The forest of the righteous] (Warszawa: Książka I Wiedza, 1968), p. 33; Shmuel Krakowski and Yisrael Gutman, *Jews and Poles during World War II* (tentative title) (New York: Holocaust Library, in press).

5. Of the 189 rescuers only 13 percent participated exclusively in organizational rescue.

6. Of all the rescuers, 75 percent had at one time or another escorted Jews to

safety. In contrast, only 40 percent of the survivors say that their helpers escorted them to safety.

7. Sixty-eight percent of the Poles offered food or money. The survivors claim that 63 percent of their helpers gave them the same kind of aid. Money, of course, could be easily transformed into food.

8. Only 4 percent of the rescuers express "negative" or "predominantly negative" attitudes toward Jews they kept at home. More importantly, however, 86 percent of them express "positive" or "predominantly positive" attitudes toward those they protected. For the rest it is hard to establish whether their help involved keeping the fugitives at home.

9. Of the rescuers, 60 percent are "very satisfied" about the help they offered, while another 36 percent are "satisfied" and 4 percent have some regrets. In this last group only two individuals say that they would not do it again.

10. Nathan Gross, "Research into the Question of Aryan Papers," *Yediot Yad Vashem,* no. 10–11 (1956): 34; Nathan Gross, "Unlucky Clara," *Yad Vashem Bulletin,* no. 15 (August 1964): 55–60. Clara had to change places twenty-five times, the known scientist Hirszfeld, eleven times. See Ludwik Hirszfeld, *Historia Jednego Życia* [The story of one life] (Warszawa: Pax, 1957).

11. Of the rescuers that kept Jews at home 58 percent say that they continued to keep them even when faced with special dangers. This figure becomes substantially lower when survivors report whether the rescuers kept them on when faced with special danger. Thirty-six percent of the survivors say that they stayed on. More specifically, the survivors report that sometimes they insisted that it was safer to leave. When looked at in terms of who made the decision, in 19 percent of the cases the fugitives left on their own insistence; in another 21 percent of the cases their departure was prompted by mutual consent.

12. Datner, *Forest of the Righteous;* Krakowski and Gutman, *Jews and Poles.*

13.

Duration of aid as reported by rescuers	Percent
Single acts*	2
Intermittent	33
Up to six months	4
Six months and over	61
	n = 189

Duration of aid as reported by survivors	Percent
Single acts	15
Intermittent	37
Up to six months	27
Six months and over	21
	n = 426

*This refers to single acts performed for different individuals.

14. Personal communication, Dr. Mordechai Paldiel, director of the Department for the Righteous, Yad Vashem, Jerusalem, Israel.

15. Some of those who feel that the number of Polish rescuers runs into hundreds of thousands are: Władysław Bartoszewski and Zofia Lewinówna, *The Samari-*

tans: Heroes of the Holocaust (New York: Twyne Publishers, Inc., 1970), p. 51; Datner, *Forest of the Righteous*, p. 33; Kazimierz Iranek-Osmecki, *He Who Saves One Life* (New York: Crown Publishers, Inc., 1971), p. 1333. The highest estimate is offered by Iranek-Osmecki, who says that "it is no exaggeration to say that to save 40,000 Jews more than a million Poles had to risk their lives."

16. The figure of 343 cited by Datner, *Forest of the Righteous*, pp. 80–85, covers the period until 1967. The figure of 554 includes only those who were identified by name and covers the period until 1968. See Władysław Bartoszewski and Zofia Lewinówna, *Ten Jest Z Ojczyzny Mojej* [This one is from my country] (Kraków: Wydawnictwo Znak, 1969), pp. 832–874. Finally, the figure of 668 includes those who were identified by name only and was published in 1971. See Stanisław Wroński and Maria Zwolakowa, *Polacy I Żydzi, 1939–1945* [Poles and Jews] (Warszawa: Książka I Wiedza, 1971), pp. 405–448.

17. Cases of burned villages have been reported by Wroński and Zwolakowa, *Poles and Jews;* see also Zvi Weigler, "Two Villages Razed for Extending Help to Jews," *Yad Vashem Bulletin*, no. 1 (1957): 18–20.

Chapter 5

1. Here are only a few of the sources where such statements are made: Shmuel Krakowski and Yisrael Gutman, *Jews and Poles during World War II* (New York: Holocaust Library, in press). Kazimierz Iranek-Osmecki, *He Who Saves One Life* (New York: Crown Publishers, Inc., 1971), p. 128; Vladka Meed, *On Both Sides of the Wall* (New York: Ghetto Fighters' House and Hakibbutz Hameuhad Publishing House, 1972), p. 245; Emmanuel Ringelblum, *Polish–Jewish Relations during the Second World War* (Jerusalem: Yad Vashem, 1974), p. 226.

2.

Rescuers' relationship to money	Percent
Some money received but not basis for aid	9
Jews contributed money to their own upkeep	23
Jews did not contribute money; rescuers supported them	52
Organizational aid—no money involved	16
	$n = 182$

3.

Survivors' reports of payment	Paid helpers (%)	Rescuers (%)
Payment sole basis for help	100	0
Some payment offered but money not important	0	9
No payment; helper expected payment later	0	1
No payment; Jew contributed to own upkeep	0	4
No payment; helper supported the Jew	0	52
No payment; hard to establish if helper supported the Jew	0	34
	$n = 86*$	$n = 427$

*The 308 survivors in the sample mentioned 565 Poles who helped them. These 86 paid helpers make up 16 percent of 535 Poles about whom the survivors offer information on payment. Included in this group are 5 percent of survivors who felt that they survived on their own.

4. Seventy-three percent of the survivors said that they asked their paid helpers for aid and that only 1 percent, or one of these Poles, made the offer first. Of those saved by righteous Christians, 62 percent approached their helpers and 15 percent were approached by these Christians.

5.

Type of aid	Paid rescue (%)	Rescue (%)
Promised help, went back on promise and robbed	17	0
Promised help, went back on promise	15	3
Promised help, kept promise	11	13
No promise of help, received help	57	84
	n = 84	n = 420

6. The numbers on which the following percentages are based vary because the extent of missing information in each category varies.

Socioeconomic status	Paid helpers (%)	Rescuers (%)
Poor and very poor	85	47
Poorly educated	90	53
Making a living as a peasant	66	42

7.

Duration of paid help	Percent
Single acts	15
Intermittent	7
Up to six months	43
Six months and over	35
	n = 86

8. The survivors said that of their paid helpers, 87 percent had offered shelter, compared to 65 percent of those saved by the righteous. Both of these figures are based on the survivors' reports; more direct information from the righteous themselves indicates that 87 percent of them offered shelter.

9. In my sample, 79 percent of the survivors said that their paid helpers escorted them to safety, while 69 percent said that these helpers warned them about danger.

10. Rosa Pinczewska, "Rosa's Journey," pp. 297–299 in *The Root and the Bough*, ed. Leo W. Schwarz (New York: Rinehardt and Co., Inc., 1949).

11. *Type of help by treatment*

Survivors' reports of treatment	Paid helpers (%)	Rescuers (%)
Good and excellent	33	96
Poor, not specified	9	2
Starved	13	1
Increases price, starves, robs	42	0
Does not apply, survived on own	3	0
	n = 76	n = 415

12. *Reactions of Jewish survivors who were protected by paid helpers* and by righteous Poles*

Reactions of survivors	Paid helpers (%)	Righteous Poles (%)
See them as courageous or good-natured	18	95
See them as "greedy" and "money hungry"	66	0
Overall negative attitudes	65	2

*The numbers on which these figures are based vary because the proportions of "no answer" vary as well.

13.

Survivors' reports of reactions	Paid helpers (%)	Rescuers (%)
Demand for more money	21	0
Demand for more money and threats to denounce	14	0
Actual denouncement after received all money	12	0
No demands for money and no threats but Jews fear that will denounce	19	1
No demands for more money and no likelihood of denouncement	34	99
	n = 78	n = 417

14. Here are just a few examples where such reactions occurred: Janka Hescheles, *Oczyma Dwunnastoltniej Dziewczyny* [With the eyes of a twelve-year-old] (Kraków: Jewish Historical Commission, 1946), p. 37; Aryeh Kolnicki, *The Diary of Adam's Father* (Israel: Ghetto Fighters' House, 1973), p. 25; Howard Roiter, *Voices from the Holocaust* (Givataim, Israel: Asurno Press, 1975), p. 92; Leon Wells, *The Janowska Road* (New York: The Macmillan Co., 1963), p. 111.

15. Wells, *Janowska Road*, p. 259.

16. Oskar Pinkus, *The House of Ashes* (New York: The World Publishing House Co., 1964), pp. 161–162.

17. *Attitude toward continuing relationship as reported by survivors*

	Paid helpers (%)	Rescuers (%)
No desire to be rid of charges	14	58
Got rid of charges*	51	39
Unfulfilled desire to be rid of charges	35	3
	n = 69	n = 322

*Some of these reactions were based on realistic external perils.

18. Zantea Margules, "Moje Życie W Tarnpolu Podczas Wojny" [My life in Tarnopol during the war], *Biuletyn Żydowskiego Instytutu Historycznego*, no. 36 (1960): 62–94; Nechama Tec, *Dry Tears: The Story of a Lost Childhood* (New York: Oxford University Press, 1984).

19. S.P. Altmann, "Simmel's Philosophy of Money," *American Journal of Sociology* 9 (1903): 46–68; Georg Simmel, *The Philosophy of Money* (Boston: Routledge and Kegan Paul, 1978).

20. Comparing commercial blood donors and blood offered free of charge, Titmuss came up with very similar conclusions. See Richard M. Titmuss, *The Gift Relationship: From Human Blood to Social Policy* (New York: Pantheon Books, 1971). See especially Chapter 14, the summary chapter.

Chapter 6

1. A few such published examples are as follows: Władysław Bartoszewski and Zofia Lewinówna, *The Samaritans: Heroes of the Holocaust* (New York: Twyne Publishers, Inc., 1970), p. 58; Philip Friedman, *Their Brothers' Keepers* (New York: Holocaust Library, 1978), pp. 114–155; Ludwik Hirszfeld, *Historia Jednego Życia* [The story of one life] (Warszawa: Pax, 1957), p. 306; Emmanuel Ringelblum, *Polish–Jewish Relations during the Second World War* (Jerusalem: Yad Vashem, 1974), p. 197.

2. Tadeusz Hołuj, "Jan Mosdorf's Truce," pp. 118–124 in *Lest We Forget*, ed. Adolf Rudnicki, (Warsaw: Polonia Foreign Languages Publishing House, 1955).

3. Emmanuel Ringelblum, *Notes from The Warsaw Ghetto* (New York: Schocken Books, 1975), p. 36; Ringelblum, *Polish–Jewish Relations*, p. 197.

4. Władysław Bartoszewski, and Zofia Lewinówna, *Ten Jest Z Ojczyzny Mojej* [This one is from my country] (Kraków: Wydawnictwo Znak, 1969), pp. 816–819.

5. The names that reappear most frequently are Monsignor Marceli Godlewski, Franciszek Kowalski, Jan Mosdorf, Leon Nowodworski, and Witold Rudnicki.

6. They made up a 60 percent majority. See Antony Polonsky, *Politics in Independent Poland, 1921–1939* (Oxford: Clarendon Press, 1972), p. 325.

7.

Rescuers' opinions about Jews in general*	Percent
Negative	15
Moderate	38
Positive	47
	n = 87

*This information is missing for 102 rescuers.

Rescuers' stereotypes about Jews*	Percent
Has many	8
Has few	27
No stereotypes	35
	n = 73

*This kind of information is missing for 116 rescuers.

8.

Survivors' reports about rescuers' attitudes toward Jews	Paid rescuers (%)	Rescuers (%)
Negative	70	26
Mixed	18	18
Positive and neutral	12	56
	n = 69	n = 351

9. I am using the real name because Vera wanted Komornicka to be remembered.

10. Michael Zylberberg, *A Warsaw Diary, 1939–1945* (London: Vallentine, Mitchell, 1969), pp. 132–133.

11. Jerzy Andrzejewski, *Martwa Fala* [Dead wave] (Warszawa: Spółdzielnia Wydawnicza, 1947), p. 39.

12. Included here is the original version of "The Protest," and a translation of the entire document (translated by my student, Doris Grinberg).

13. Joseph Kermish, "The Activities of the Council for Aid to Jews (Żegota) in Occupied Poland," pp. 367–398 in *Rescue Attempts during the Holocaust*, eds. Yisrael Gutman and Efraim Zuroff (Jerusalem: Yad Vashem, 1977).

14. Zofia Kossak-Szczucka, "Komu Pomagamy?" [Whom do we help?], *Prawda, Pismo Frontu Odrodzenia*, August–September 1943, p. 8 (translated from the Polish).

Chapter 7

1. Moshe Bejski, "The Righteous among the Nations and Their Part in the Rescue of Jews," pp. 582–607 in *The Catastrophe of European Jewry*, eds. Yisrael Gutman and Livia Rothkirchen (Jerusalem: Yad Vashem, 1976).

2. Tatiana Berenstein and Adam Rutkowski, "O Ratowaniu Żydow Przez Polaków W Okresie Okupacji Hitlerowskiej," [Polish rescue of Jews during the occupation], *Biuletyn Żydowskiego Instytutu Historycznego*, no. 35 (1960): 29; Bernard Goldstein, *The Stars Bear Witness* (New York: The Viking Press, 1949), p. 232; Kazimierz Iranek-Osmecki, *He Who Saves One Life* (New York: Crown Publishers, Inc., 1971), pp. 263–280.

3. Arieh Bauminger, "Righteous Gentiles," *Jewish Spectator* 27–31 (September 1964): 9–10.

4. Bejski, "The Righteous among the Nations," p. 592; Berenstein and Rutkowski, "Polish Rescue of Jews," p. 28; Emmanuel Ringelblum, *Polish–Jewish Relations during the Second World War* (Jerusalem: Yad Vashem, 1974), p. 199.

5. Of the survivors, 21% identify their rescuers as working class. Included in this category are skilled and unskilled laborers, servants, governesses.

6. Of the rescuers, 22 percent can be identified as working class, using the same as criteria as in note 5 above.

7. Existing categories and class divisions tend to fluctuate. Some analysts identify the following classes among Poland's prewar population: (1) intelligentsia, (2) workers, (3) peasants and marginal classes, and (4) burghers and landowners; others use the categories of (1) peasants and agricultural workers, (2) industrial workers, (3) entrepreneurs, and (4) intelligentsia.

Social classes in prewar Poland	Distribution (%)
Higher classes (intelligentsia, intellectuals, etc.)	10
Middle class (business people, entrepreneurs, etc.)	5
Lower class (laborers)	20
Peasants	60
Miscellaneous	5

The accompanying tabulation is based on these and other similar classifications, and on the following sources: Alicja Iwanska, *Contemporary Poland: Society, Politics, Economy* (Chicago: The Chicago University Press, 1955), pp. 100–136; Antony Polonsky, *Politics in Independent Poland, 1921–1939* (Oxford: Clarendon Press, 1972), pp. 24–35; Edward D. Wynot, Jr., *Polish Politics in Transition* (Athens, Ga.: University of Georgia Press, 1974), p. 7.

8. A few of these sources are as follows: Tatiana Berenstein and Adam Rutkowski, *Assistance to the Jews in Poland, 1930–1945* (Warsaw: Polonia Foreign Languages Publishing House, 1963), pp. 37–38; Janina Dumin-Wąsowicz, "Wspomnienia O Akcji Pomocy Żydom Podczas Okupacji Hitlerowskiej" [Recalling aid to Jews during occupation]. *Biuletyn Żydowskiego Instytutu Historycznego* no. 45–46 (1963): 248–261; Ringelblum, *Polish–Jewish Relations,* p. 199.

9. This category is loosely defined, including such professionals as writers, painters, journalists, actors, physicians, engineers, teachers, high civil servants, school directors.

10. Using the same criteria as in the Jewish sample (note 9), 54 percent of the rescuers can be classified as intelligentsia or intellectuals.

11. Thus, the estimated proportion of intellectuals in the population at large was 10 percent, as compared to 19 percent of intellectuals among the rescuers that were mentioned by the Jewish sample.

12. Ludwik Hirszfeld, *Historia Jednego Życia* [The story of one life] (Warszawa: Pax, 1957), p. 356.

13. Zygmunt Klukowski, *Dziennik Z Lat Okupacji, Zamojszczyzny* [Wartime diary from Zamość] (Lublin: Lubelska Spółdzielnia Wydawnicza, 1958), p. 109; Aryeh Kolnicki, *The Diary of Adam's Father* (Israel: Ghetto Fighters' House, 1973). The author of the diary refers repeatedly to the theme of peasants and their cruelty to the Jews; Stanisław Ziemiński, "Kartki Dziennika Nauczyciela W Łukowie Z Okresu Okupacji Hitlerowskiej" [Notes from a teacher's diary from Łukow during Hitler's occupation], *Biuletyn Żydowskiego Instytutu Historycznego,* no. 27 (1958): 109.

14. Among the rescuers the proportion of peasants dwindles to 12 percent. As with the intelligentsia, this small percentage may be due to a sample bias.

15. Ringelblum, *Polish–Jewish Relations,* p. 148.

16. Ibid., p. 198.

17. Recall that in prewar Poland 5 percent of the population were middle class; this is scarcely different from 3 and 4 percent.

18. Information from Jewish survivors about their rescuers' motivations for help is incomplete; such information was available only for 89 rescuers, or fewer than 25 percent of rescuers.

19. Survivors fail to touch on underground participation or political involvement of paid helpers. Thus, lack of information eliminates profit helpers from all discussion of politics in this regard.

20. For discussions about the positions toward Jews of the political Left and Right see Reuben Ainsztein, "The Jews of Poland Need Not Have Died," *Midstream,* Autumn 1958, pp. 2–4, 101–103; Berenstein and Rutkowski, "Polish Rescue of Jews," pp. 27–28; Celia S. Heller, *On the Edge of Destruction* (New York: Columbia University Press, 1977), pp. 85–130; Ringelblum, *Polish–Jewish Relations,* p. 198; Bernard D. Weinryb, "The Jews in Poland," pp. 207–326 in *The Jews in the Soviet Satellite,* eds. Peter Meyer, et al. (Westport, Conn.: Greenwood Press, 1953); Edward D. Wynot, Jr., "A Necessary Cruelty: The Emergence of

Official Anti-Semitism in Poland, 1936–1939," *The American Historical Review* 76, no. 4 (October 1971): 1035–1058.

21. Philip Friedman, *Their Brothers' Keepers*, (New York: The Holocaust Library 1978), p. 127.

22. All publications on Poland's prewar and wartime political life emphasize political diversity. For a few such references see note 20.

23. For some discussions on this topic see Reuben Ainsztein, *Jewish Resistance in Nazi-Occupied Eastern Europe* (New York: Barnes and Noble, 1974), p. 588; Marek Arczyński and Wiesław Balcerak, *Kryptonim Żegota* [Żegota] Warszawa: Czytelnik, 1979); Władysław Bartoszewski, *The Blood Shed Unites Us* (Warsaw: Interpress Publishers, 1970), p. 228; Berenstein and Rutkowski, "Polish Rescue of Jews," p. 37; Friedman, *Their Brothers' Keepers*, p. 120; Teresa Prekerowa, *Konspiracyjna Rada Pomocy, Żydom W Warszawie 1942–1945* [Illegal Help to Jews] (Warszawa, Państwowy Instytut Wydawniczy, 1982); Ringelblum, *Polish–Jewish Relations*, p. 293.

24. Marek Arczyński, "Rada Pomocy Zydom W Polsce" [Council for Aid to Jews], *Biuletyn Żydowskiego Instytutu Historycznego*, no. 65–66 (1968): 173–185; Prekerowa, *Illegal Help*, pp. 66–107.

25. According to one reliable source, it is questionable whether the funds destined for the Jews in Poland reached them. The same source also points out that the Polish underground allotted less than 1 percent of the $30 million it had received from abroad for helping Jews. See Emmanuel Ringelblum, *Polish–Jewish Relations* (Jerusalem: Yad Vashem, 1974), p. 213. In contrast, Prekerowa in her discussion about funds emphasizes the care with which members of the Żegota distributed money. See Prekerowa, *Illegal Help*, pp. 108–146. Whether and to what extent the government in exile and the underground in Poland contributed adequately to the support of Jews is a still much-debated question. Most agree that whatever sums reached members of Żegota, these were distributed honestly and with care.

26. Official financial aid to Jews remains a controversial topic. Without taking a position about the adequacy of such funds, the existing evidence points to overall budgetary expansions of Żegota. Not surprisingly, these came about with organizational expansion. For details see Kermish, "Activities of the Council for Aid to Jews," pp. 367–398. In 1943, for example, the monthly sum received amounted to 150,000 złotys; in 1944 it reached the 2 million złotys level. These figures are cited in Prekerowa, *Illegal Help*, pp. 114–116.

27. These names and their corresponding positions reappear in most of the publications cited here (e.g., Arczyński, Kermish, Prekerowa).

28. Kermish, "Activities of the Council for Aid to Jews," p. 372.

29. Of the helpers, 17 percent say that they attended to wounded Jews. Of the survivors, 6 percent say that they received medical aid from Poles. In addition, 53 percent of the rescuers supplied the fugitives with forged documents. Even before the Council for Aid to Jews came into existence, segments of the Polish underground had supplied Jews with forged documents. See Edward Chodźiński, "Pomoc Żydom Udzielana Przez Konspiracyjne Biuro Fałszywych Dokumentów W Okresie Okupacji Hitlerowskiej" [Help offered to Jews during the war by the underground office of illegal documents], *Biuletyn Żydowskiego Instytutu Historycznego*, no. 75 (1970): 129–132. The author emphasizes that the offering of false documents to Jews was approved of by the Delegatura.

30. Prekerowa, *Illegal Help*, pp. 189–217, 478.

31. Jan Karski, *Story of a Secret State* (Boston: Houghton Mifflin Co., 1944);

Stefan Korboński, *The Polish Underground State: A Guide to the Underground, 1939–1945* (Boulder, Colo.: East European Quarterly, 1978).

32. Paul Lendvai, *Anti-Semitism without Jews* (New York: Doubleday, 1971), p. 56.

33. The divisiveness of the Polish underground is well documented. For a few examples see Kazimierz Iranek-Osmecki, *He Who Saves One Life* (New York: Crown Publishers, Inc., 1971); Tadeusz Komarowski-Bor, *The Secret Army* (London: Victor Gallancz, Ltd., 1950); Jan Nowak, *Courier from Warsaw* (Detroit, Mich.: Wayne State University Press, 1982); Wacław Zagórski, *Wolność W Niewoli* [Freedom in slavery] (London: Nakładem Autora I Czytelników Przedpłacicieli, 1971).

34. A document marked No. 171/43, drafted at the end of 1943 by the Home Army H.Q. Information and Propaganda Office was included by Jósef Kermish and Shmuel Krakowski, who edited and footnoted Ringelblum's book. See Ringelblum, *Polish–Jewish Relations*, pp. 223–225. Also see Wacław Zagórski, [Freedom in Slavery], p. 310.

35. The following is but a small sample of contradictory positions about the political direction of organized help and its inadequacy or adequacy: Mordechai Anielewicz, "The Jewish Fighter's Organization Appeals," *Extermination and Resistance, Historical Records and Source Material* 1 (1958): 7–9; Ainsztein, *Jewish Resistance*; Ainsztein, "The Jews of Poland Need Not Have Died"; Chodziński, "Help offered to Jews during the war by the underground office of illegal documents." Iranek-Osmecki, *He Who Saves One Life*; Komarowski-Bor, *The Secret Army*; Hanna Krall, *Zdążyć Przed Panem Bogiem* [To make it before God] (Kraków: Wydawnictwo Literackie, 1977); Joseph L. Lichten, "Did Polish Jews Die Forsaken?" *The Polish Review* 4, no. 1–2 (1959): 119–126; Teodor Niewiadomski, "Z Pomocy Dla Warszawskiego Getta" [Help to the Warsaw ghetto] *Kultura*, Warszawa, April 21, 1968, p. 6; S.W. Pollack, "Need They Have Died?", *World Jewry* (November 1958): 14–15; "The JFO and the Polish Underground," *Extermination and Resistance, Historical records and Source Material* 1 (1958): 14–19.

36. Julian Eugeniusz Kulski, *Dying We Live* (New York: Holt, Rinehart and Winston, 1980), pp. 133–136.

37. Berenstein and Rutkowski, "Polish Rescue of Jews" p. 44; Ainsztein, *Jewish Resistance*, p. 638.

38. Even though Ainsztein emphasizes that the political Left was aiding Jews, he cites a number of operations that were initiated by the Home Army's political Center or right of the center. See Ainsztein, *Jewish Resistance*, pp. 638–642.

39. For a few additional descriptions of help to the Warsaw ghetto uprising by nonleftist segments of the AK see Szymon Datner, *Las Sprawiedliwych*, [The forest of the righteous] (Warszawa: Książka I Wiedza, 1968), p. 96; Goldstein, *The Stars Bear Witness*, p. 178; Balicka Helena Kozłowska, *Mur Miał Dwie Strony* [The wall had two sides] (Warszawa: Wydawnictwo Ministerstwa Obrony Narodowej, 1958), p. 88; Ringelblum, *Polish–Jewish Relations*, p. 179.

40. Iwański's case is described in many publications, and both historians and rescuers referred to him in private conversations. For a few such references see Ainsztein, *Jewish Resistance*, p. 647; Bejski, The Righteous among the Nations, p. 96; Iranek-Osmecki, *He Who Saves One Life*, p. 164; Henryk Iwański, "Mówi Major Bystry" [Major Bystry speaks], *Kultura*, Warszawa, April 21, 1968, p. 6; Dan Kurzman, *The Bravest Battle: The Twenty-Eight Days of the Warsaw Ghetto Uprising* (New York, G.P. Putnam and Sons, 1976), pp. 287–288.

41. Kazimierz Moczarski, *Rozmowy Z Katem* [Talks with a hangman] (Warszawa: Państwowy Instytut Wydawniczy, 1978).

42. Sixty percent of the rescuers fit into the "no political affiliation" category. This figure is based on 157 individuals for whom information about political affiliation is available.

43. The following tabulation considers only rescuers who expressed definite political preferences.

Political preferences of rescuers	Percent
Communists and Socialists	60
Social Democrats and Peasants	30
Catholics and National Democrats	10
	$n = 63$

44. The following distribution of survivors' reports about their rescuers' political leanings refers only to those rescuers about whom some information about political leanings was available.

Rescuers' political leanings as reported by survivors	Percent
Communists and Socialists	21
Social Democrats and Peasants	3
Catholics, National Democrats, Sanacja	4
Patriot without specification	72
	$n = 110$

45. The following distribution considers only those rescuers who were described in terms of their specific political leanings.

Rescuers' specific political leanings as reported by survivors	Percent
Communists and Socialists	74
Social Democrats and Peasants	10
Catholics and National Democrats	16
	$n = 31$

46. See Polonsky, *Politics in Independent Poland*, p. 325.

Chapter 8

1. Mary Berg, *Warsaw Ghetto: A Diary* (New York: L.B. Fischer, 1945), p. 210; Kazimierz Iranek-Osmecki, *He Who Saves One Life* (New York: Crown Publishers, Inc., 1971), p. 128.

2. Moshe Bejski, "The Righteous among the Nations and Their Part in the Rescue of Jews," p. 592 in *The Catastrophe of European Jewry*, eds. Yisrael Gutman and Livia Rothkirchen (Jerusalem: Yad Vashem, 1976).

3. Emmanuel Ringelblum, *Polish–Jewish Relations during the Second World War* (Jerusalem: Yad Vashem, 1974), p. 96; Henryk Grynberg, *Childhood of Shad-*

ows (London: Vallentine, Mitchell, 1969), pp. 79–82; Michael Zylberberg, *A Warsaw Diary, 1939–1945* (London: Vallentine, Mitchell, 1969), pp. 106–107.

4. Rescuer–Rescued Relationship before the Beginning
of Rescue as Reported by Jewish Survivors:

Relationship	Percent
Strangers	51
Acquaintances	21
Work with commercial ties	9
Friends	19
	$n = 412$

5.

Rescuers' ties to Jews	Percent
Familial* and friendship	55
Commercial	15
Working	10
No ties	20
	$n = 145$

*Familial ties were practically nonexistent and as such referred to not more than three cases.

6.

Rescuers' ties to Jews as reported by survivors	Percent
Familial* and friendship	47
Commercial	24
Working	14
No ties	15
	$n = 135$

*Here too familial ties were practically nonexistent, referring to two cases.

7.

Paid helpers' ties to Jews	Percent
Familial and friendship	0
Commercial	53
Working	18
No ties	29
	$n = 17*$

*Because the number of cases is small these results should be viewed with caution.

8. Janka Hescheles, *Oczyma Dwynastoltniej Dziewczyny* [With the eyes of a twelve-year-old] (Kraców: Jewish Historical Commission, 1946), pp. 43–55.

9. M.J. Feigenbaum, "Life in a Bunker," p. 146 in *The Root and the Bough*, ed. Leo W. Schwarz (New York: Rinehart and Co., Inc., 1949).

10. Leon Wells, *The Janowska Road* (New York: The Macmillan Co., 1963), pp. 224–227.

11. Forty-three percent of the rescuers offered help to a variety of people: strangers, acquaintances, and friends.

Chapter 9

1. Howard Roiter, *Voices from the Holocaust* (Givataim, Israel: Asurno Press, 1975), p. 170.

2. Tadeusz Hołuj, "Jan Mosdorf's Truce," pp. 118–124 in *Lest We Forget*, ed. Adolf Rudnicki (Warsaw: Polonia Foreign Languages Publishing House, 1955), p. 119.

3. Jerzy Pfefer, "Moja Ucieczka Z Majdanka" [My escape from Majdanek], *Biuletyn Żydowskiego Historycznego*, No. 71–72, 1969, p. 218.

4. Leon Wells, *The Janowska Road* (New York: The Macmillan Co., 1963), p. 257.

5. The Vatican denies free access to its World War II archives, but this has not prevented some scholars from studying the Church's position toward the Nazi extermination of Jews. For a few examples see Yehuda Bauer, *The Holocaust in Historical Perspective* (Seattle: University of Washington Press, 1978); Saul Friedlander, *Pius XII and the Third Reich: A Documentation* (New York: Octagon Books, 1980); Walter Laquer, *The Terrible Secret: Suppression of the Truth about Hitler's Final Solution* (Boston: Little, Brown and Co., 1980); John F. Morley, *Vatican Diplomacy and the Jews during the Holocaust, 1939–1943* (New York: KTAV Publishing House, Inc., 1980).

6. I agree with the historian Philip Friedman that when it comes to the Catholic clergy and aid to Jews we must distinguish between the official policy of the Vatican and informal reactions by clergy. Even though Pope Pius XII never officially condemned the Nazi extermination of Jews, he did not object to informal rescue attempts by clergy. Unofficially the Pope himself helped a few prominent Jewish individuals. In 1944, with Hitler clearly defeated, Pius XII urged the Hungarian government to halt Jewish death camp deportations. See Philip Friedman, "Righteous Gentiles in the Nazi Era," pp. 409–421 in *Roads to Extinction: Essays on the Holocaust*, ed. Ada June Friedman (Philadelphia: The Jewish Publications Society of America, 1980).

7. To this day theologians debate the Holocaust and its relation to Christianity. To some, the Nazi extermination of Jews represents a threat to Christianity. See a collection of essays on the topic written by different theologians: Harry James Cargas, ed., *When God and Man Failed* (New York: The Macmillan Co., 1981); see also Harry James Cargas, *A Christian Response to the Holocaust* (Denver, Colo: Stonehenge Books, 1981).

8. Arieh Sutzkever, "Never Say This Is the Last Road," pp. 66–92 in *The Root and the Bough* ed. Leo, W. Schwarz, (New York: Rinehart and Co. Inc., 1949).

9. Nathan Gross, "Aryan Papers in Poland," *Extermination and Resistance, Historical Sources and Material* 1 (1958): 79–86; Hanna Krall, *Zdążyć Przed Panem Bogiem* [To make it before God] (Kraków: Wydawnictwo Literackie, 1977).

10. Emmanuel Ringelblum, *Polish–Jewish Relations during the Second World War* (Jerusalem: Yad Vashem, 1974), pp. 208–209. Some of the others who have pointed to the clergy's rescue of Jews are as follows: *Chrzescijańskie Stowarzyszenie Społeczne* [Christian charity] Dzieło Miłosierdzia Chrzescijańskiego, (Warszawa: 1968), pp. 14–15; Szymon Datner, "Materiały Z Dźiedźiny Ratownictwa Żydów W Pólsce W Okresie Okupacji Hitlerowskiej" [Materials about Jewish rescue during the occupation] *Biuletyn Żydowskiego Instytutu Historycznego*, no. 73 (1970): 135–137; Philip Friedman, *Their Brothers' Keepers* (New York: Holocaust Library, 1978), pp. 21–32.

11. Except for one nun, none of the profit helpers were identified as clergy. This one case, as will be seen shortly, refers to the nuns who protected Karla Mintz.

12. These percentages would be more meaningful if they could be compared to the proportion of clergy in the population at large. As a rule, occupational breakdowns do not list the clergy as a special category. All we know is that in Poland at the outbreak of World War II there were 16,000 priests, 4500 lay brothers, and 17,000 sisters. *New Catholic Encyclopedia*, vol. 2 (New York: McGraw-Hill, 1967), pp. 481–483. A more detailed breakdown essentially reconfirming these figures is offered by Jerzy Smolewski, "Ludność," pp. 21–27 in *Encyclopedia Polska* (Kraków: Wydawnictwo Gutenberg, 1931).

13. See *Christian Charity*, p. 17; Datner, "Materials about Jewish Rescue," pp. 133–138; Władysław Smolski, *Losy Dziecka* [Child's fate] (Warszawa: Pax, 1961), p. 241. Herself saved by nuns, Zofia Szymańska focuses on aid by convents, particularly as offered to children. A few of her publications are *Byłam Tylko Lekarzem* [I was only a physician] (Warszawa: Instytut Wydawniczy Pax, 1979); "Ratunek W Klasztorze" [Rescue in the convent], *Biuletyn Żydowskiego Instytutu Historycznego* 4, no. 88 (1973): 33–44; "Ze Wspomień Lekarza" [From a physician's memories], *Biuletyn Żydowskiego Instytutu Historycznego* 4, no. 80 (1971): 51–64.

14. Ringelblum, *Polish–Jewish Relations*, pp. 208–209.

15. Janina David, *A Touch of Earth: A Wartime Childhood* (New York: Orion Press, 1969), p. 139.

16. Alexander Donat, *The Holocaust Kingdom: A Memoir* (New York: Holt, Rinehart and Winston, 1965), p. 350.

17. For one such touchingly told story, see Peter Hellman, *Avenue of the Righteous* (New York: Atheneum Publishers, 1980), pp. 168–267.

18. Of the 116 survivors, 72 percent said that their helpers were "very religious," 20 percent said that they were "not religious," and the rest were "moderately religious." These figures cannot be compared to paid helpers because information about their religiosity is available for only nine individuals.

19. Of the rescuers, 49 percent are "very religious," 11 percent "moderately religious," and 40 percent "not religious."

20. Case no. 206, Hirshaut Archival Collection, YIVO, Institute for Jewish Research, New York City.

21. Case no. 39, Pomoc Polaków, 1–180, Archival Collection, Jewish Historical Institute in Warsaw, Warsaw.

Chapter 10

1. Perry London, "The Rescuers: Motivational Hypotheses about Christians Who Saved Jews from the Nazis," pp. 241–250 in Jacqueline R. Macaulay and Leonard Berkowitz, eds., *Altruism and Helping Behavior* (New York: Academic Press, 1970).

2. Ibid.; Douglas Huneke, "A Study of Christians Who Rescued Jews during the Nazi Era," *Humboldt Journal of Social Relations* 9, no. 1 (1981–1982): 145–150; Samuel P. Oliner, "The Heroes of the Nazi Era: A Plea for Recognition," *Reconstructionist* 48 (June 1982): 7–14.

3. Anna Freud, "A Form of Altruism," pp. 132–148 in *The Ego and the Mechanism of Defense*, ed. Anna Freud (New York: International Universities Press. Inc., 1946).

4. Peter M. Blau, "Social Exchange," pp. 452–457 in *International Encyclope-*

dia of the Social Sciences, ed. David L. Sills (New York: The Macmillan Co., 1968); Alvin W. Gouldner, "The Norm of Reciprocity: A Preliminary Statement," *American Sociological Review* 25, no. 2 (April 1960): 161–178; Georg Simmel, pp. 379–395 in *The Sociology of Georg Simmel,* ed. Kurt H. Wolff (Glencoe, Ill.: The Free Press, 1950).

5. Blau, "Social Exchange"; Pitrim A. Sorokin, "Factors of Altruism and Egoism," *Social Science Research* 32 (1948): 674–678; Lauren Wispe, "Toward an Integration," pp. 303–327 *Sympathy and Helping: Psychological and Sociological Principles,* ed. Lauren Wispe (New York: Academic Press, 1978).

6. Edward O. Wilson, "The Genetic Evolution of Altruism," p. 11 in Wispe, ed., *Sympathy and Helping.*

7. These are but a few examples of bystander research. L. Bickman, "Effects of Another Bystander's Ability to Help in Bystander Intervention in an Emergency," *Journal of Experimental Social Psychology* 6 (May 1971): 367–379; L. Bickman, "Social Influence and Diffusion of Responsibility," *Journal of Experimental Social Psychology* 8, no. 5 (September 1972): 438–445; Arline R. Brenner and James M. Levin, "Off-Duty Policemen and Bystander Apathy," *Journal of Police Science and Administration* 1, no. 1 (March 1973): 61–64; John Darley and Bibb Latane, "When Will People Help?" *Psychology Today* 2, no. 7 (1968): 54, 56–57, 70–71; James C. Hackler, Ho Kwai-Yiu, and Carol Urquhart-Ross, "The Willingness to Intervene," *Social Problems* 21, no. 3 (1974): 328–344; Bibb Latane and Judith Rodin, "A Lady in Distress: Inhibiting Effects of Friends and Strangers on Bystander Intervention," *Journal of Experimental Social Psychology* 5 (1969): 189–202; Jane Allyn Piliavin, et al., "Responsive Bystanders: The Process of Intervention," pp. 279–304 in *Cooperation and Helping Behavior Theories and Research,* eds. Valerian J. Derlega and Janush Grzelak (New York: Academic Press, 1982); Wispe, "Toward an Integration."

8. In one of the relatively rare natural setting studies, Rosenhan examined a group of student freedom riders. He found that those who had continuously participated in the civil rights movement had more positive relations with their parents and their parents were more consistently committed to altruism than those who limited their civil rights participation to a single act. See Rosenhan, "Natural Socialization of Altruistic Autonomy," pp. 251–268 in *Helping Behavior,* eds. J. R. Macaulay and L. Berkowitz.

9. A study conducted in California points to a fascinating exception. California passed the so-called Samaritan law that offers financial compensation to people who when witnessing a crime try to prevent it from happening and are injured in the process. Social scientists had studied a group of such "good samaritans." The surprising and very different results showed that those who interfered with a crime were entering into a contest with the potential criminal. They were angry, often gun owners, who showed little concern for the victim. To be sure, while this interference involved serious risks it also had to do with single spontaneous acts. As such, however, this study raises some fascinating issues that I only touched on. A minority of Christian rescuers in my study felt that their protection of Jews was motivated by their desire to oppose the Nazis. See Ted L. Huston et al., "The Angry Samaritans," *Psychology Today,* June 1976, pp. 61–64, 81.

10. Leonard Berkowitz, "Social Norms, Feelings and Other Factors Affecting Helping and Altruism," pp. 63–108 in Derlega and Grzelak, eds., *Cooperation and Helping Behavior;* Dennis L. Krebs, "Altruism—An Examination of the Concept and a Review of the Literature," *Psychological Bulletin* 73 (1970): 258–302; Eliza-

beth Midlarski, "Aiding Responses: An Analysis and Review," *Merrill Palmer Quarterly* 14 (1968): 229–260; Wispe, "Toward an Integration."

11. Gordon B. Forbes et al., "Willingness to Help Strangers as a Function of Liberal Conservative or Catholic Church Membership: A Field Study with the Lost Letter Technique," *Psychological Reports* 28 (1971): 947–949; Roland Lowe and Gary Richter, "Relation of Altruism to Age, Social Class and Ethnic Identity," *Psychological Reports* 33 (1973): 567–572.

12. Jeff Ehlert et al., "The Influence of Ideological Affiliation on Helping Behavior," *The Journal of Social Psychology* 89 (1973): 315–316.

13. Mary B. Harris, "The Effects of Performing One Altruistic Act on the Likelihood of Performing Another," *Journal of Social Psychology* 88 (1972): 65–73.

14. Frank W. Schneider and Zig Mokus, "Failure to Find a Rural Urban Difference in Incidence of Altruistic Behavior," *Psychological Reports* 35 (1974): 294.

15. Leonard Bickman and Mark Kamzan, "The Effect of Race and Need on Helping Behavior," *The Journal of Social Psychology* 89 (1973): 73–77.

16. Krebs, "Altruism"; Midlarski, "Aiding Responses"; Piliavin et al., "Responsive Bystanders"; David L. Rosenhan, "Towards Resolving the Altruism Paradox: Affect, Self-Reinforcement and Cognition," pp. 101–113 in Wispe, ed., *Sympathy and Helping;* Wispe, "Toward an Integration."

17. Jacqueline R. Macaulay and Leonard Berkowitz, "Overview," pp. 1–9 in Macaulay and Berkowitz, eds., *Altruism and Helping Behavior*. I leaned on their definition of altruism.

18. This theoretical distinction has been suggested by David L. Rosenhan, "The Natural Socialization of Altruistic Autonomy," pp. 251–268 in Macaulay and Berkowitz, eds., *Altruism and Helping Behavior*.

19. London, "The Rescuers."

20. Harold Flender, *Rescue in Denmark* (New York: Simon and Schuster, 1963), p. 181.

21. Jeffrey Jaffre, *Stepson of a People* (Montreal: Admiral Printing, 1968.

22. Peter Hellman, *Avenue of the Righteous* (New York: Atheneum Publishers, 1980), pp. 3–58.

23. Philip Hallie, *Lest Innocent Blood Be Shed* (New York: Harper and Row, 1979).

24. This is the common spelling of his name. One exception is Philip Friedman, who spells it as Westerville. See *Your Brothers' Keepers* (New York: Holocaust Library, 1978), p. 66.

25. Nora Levin, *The Holocaust, 1933–1945* (New York: Schocken Books, 1973), p. 418.

26. Gideon Hausner, *Justice in Jerusalem* (New York: Holocaust Library, 1968), p. 260.

27. Donia Rosen, *The Forest My Friend* (New York: World Federation of Bergenbelson Association, 1971), p. 101.

28. The literal translation of the German word *Aktion* is "action." The Nazis applied the term to all forceful deportations of Jews, usually from the ghettoes to concentration camps.

29. P. Wizling, "Unserer Retter" [Our saviors], pp. 307–321 in *Sefer Horodenko* [The book of Horodenko], ed. S. Metzger (Israel and United States: published by former residents of Horodenko and vicinity, 1963).

30. Of the thirty-four Jewish survivors interviewed, seven were protected by paid helpers and three others described paid helpers who aided their relatives.

31. Of the entire group of survivors, fewer than one-third touched on their rescuers' integration into their milieu. Of the few who did, only 40 percent categorized them as outsiders. Absence of this kind of information supports the idea that Jewish survivors were unaware about their protectors' lack of integration. As a rule the Jewish survivors described their protectors in positive terms. It is possible that they view being an outsider negatively and therefore would refrain from describing their helpers in these terms.

32. Only 6 percent of the rescuers are dependent on the opinion of others; 47 percent are not dependent on the opinion of others, and another 47 percent are totally self-reliant.

33. The large sample contains no information about these Poles' need to follow their personal values.

34. Peter Hellman, *Avenue of the Righteous*, pp. 264–265.

35. Ruth M. Gruber, "The Pole Who Saved 32 Jews," *Hadassah Magazine*, December 1968, pp. 21, 36.

36.

Rescuers' reports of fears	Percent
Experienced great fears	22
Experienced fears, did not dwell on them	28
No fears reported	50
	$n = 168$

37. I am assuming that there is little difference between being fearful and lacking in courage. Recall that only 15 percent of the rescuers were defined as lacking in courage.

38.

Rescuers' experiences directly related to Jewish rescue	Percent
Threats and dangers (e.g., blackmail, denouncement, arrest)	80
Actual arrests, imprisonments	12
No such experiences	8
	$n = 179$

39.

Rescuers' experiences unrelated to Jewish rescue	Percent
Threats and dangers of arrest	41
Actual arrests, imprisonments	8
Actual arrests, imprisonments experienced by family of rescuer	9
No such adverse experiences	42
	$n = 168$

40. All rescuers had a hard time answering questions that dealt directly with motivations. Information about motivations comes mainly from indirect sources.

41. Reluctance to talk about motivations, together with the overall acceptance of rescue, explains in part why both the published and unpublished memoirs of these Poles contain no direct explanations for rescue.

In the few exceptional cases where such statements were made, they closely resemble those offered during my interviews. Thus, for example, a miner's wife who sheltered Jews, when pressed said: "I am like that, that's all. I cannot stand it when they do wrong to innocent people." See Władysław Bartoszewski, *The Blood Shed Unites Us* (Warsaw: Interpress Publishers, 1970), pp. 196–200. Similarly, in another instance a different rescuer is only able to say, "I did what every human being should have done." See Władysław Bartoszewski and Zofia Lewinówna, *Ten Jest Z Ojczyzny Mojej* [This one is from my country] (Kraków: Wydawnictwo Znak, 1969), pp. 431–434.

42. Indirect motivational data on rescue show that 95 percent of these helpers were prompted by Jewish need for help. This is in sharp contrast to the 26 percent who helped because it was a Christian duty or to the 52 percent who did so as a protest against the Nazi occupation. Clearly, however, more than one kind of motivation was involved.

43. Of the Jewish survivors 81 percent said that it was their suffering that made these Christians offer them help.

44. An earlier chapter showed that 51 percent of the survivors were helped by total strangers and only 19 percent by friends. The rest were by acquaintances. Consulting the sample of Polish rescuers reveals similar data. Rescuers offered help to different groups in the following percentages:

Friends only	9
Strangers only	34
Acquaintances only	13
Friends and others	47
Strangers and others	75
Acquaintances and others	45
	$n = 189$

The preceding distribution clearly shows that help offered to strangers was most common.

45. Pomoc Polaków, 1–180, Relacja 3026, Archives of The Jewish Historical Institute in Warsaw.

46. Zygmunt Klukowski, *Dziennik Z Lat Okupacji, Zamojszczyzny* [Diary from Zamość] (Lublin: Lublelska Spóldzielnia Wydawnicza, 1958).

47. Stanisław Zieminński, "Kartki Dziennika Nauczyciela W Lukowie Z Okresu Okupacji Hitlerowskiej" [Notes from a teacher's diary from Łukow during the occupation], *Biuletyn Żydowskiego Instytutu Historycznego*, no. 27 (1958): 105–112.

Chapter 11

1. Of the rescuers examined by Oliner, the single largest group were professionals, a category that corresponds to the intellectuals in my research. His information refers to righteous Christians from a variety of European countries. See Samuel P. Oliner, "The Unsung Heroes in Nazi-Occupied Europe: The Antidote for Evil," *Nationalities Papers* 12, no. 1 (Spring 1984): 129–136. Another systematic effort at examining righteous Christians, one that reported a positive relationship between

being an intellectual and becoming a rescuer, comes from Germany. See Manfred Wolfson, "Zum Widerstand gegen Hitler: Umriss eines Gruppenporträts Deutscher Retter von Juden" [Opposition to Hitler: Profile of the German rescuers of Jews]. *Tradition und Newbeginn: Internationale Forschungen Deutscher Geschichte im 20, Jahrhundert* 26 (1975): 391–407.

2. A study of kidney donors found that none of them weighed the alternatives nor did they make a conscious and rational decision to donate their organ. Protecting Jews and offering a kidney to a relative, although very different, each began in a similar way. See Carl H. Fellner, and John R. Marshall, "Kidney Donors," pp. 269–281 in Jacqueline R. Macaulay and Leonard Berkowitz, eds., *Altruism and Helping Behavior* (New York: Academic Press, 1970). This is by no means unusual when people start on important high-risk activities.

Postscript

1. This group includes Jews who passed but who, because of special circumstances, became partisans (9 percent). Moreover, it includes others who were at first passing as Poles but ended up as concentration camp inmates, and/or some who ran away from concentration camps and later on lived illegally among Christians (9 percent). Regardless of these special experiences, however, their survival mainly depended on staying illegally in the forbidden Christian world.

2. For a few attempts to examine more systematically the rescuing of Jews see Douglas Huneke, "A Study of Christians Who Saved Jews during the Nazi Era," *Humboldt Journal of Social Relations* 9, no. 1 (1981–1982): 144–150; Perry London, "The Rescuers: Motivational Hypothesis about Christians Who Saved Jews from the Nazis," pp. 241–250 in *Altruism and Helping Behavior*, eds. J. Macaulay and L. Berkowitz (New York: Academic Press, 1970), Samuel P. Oliner, "The Unsung Heroes in Nazi-Occupied Europe: The Antidote for Evil," *Nationalities Papers* 12, no. 1 (Spring 1984): 129–136.

3. For most of these cases I relied on the following collections: Władysław Bartoszewski and Zofia Lewinówna, *Ten Jest Z Ojczyzny Mojej* [This one is from my country] (Kraków: Wydawnictwo Znak, 1969); David P. Boder, *Topical Autobiographies of Displaced People*, mimeo., 16 vols. (New York: YIVO, Institute for Jewish Research, 1950–1956); Arieh L. Bauminger, *Roll of Honor* (Tel-Aviv: Hamenorah, 1970); Szymon Datner, *Las Sprawiedliwych* [The forest of the righteous] (Warszawa: Książka I Wiedza, 1968); Kazimierz Iranek-Osmecki, *He Who Saves One Life* (New York: Crown Publishers, Inc., 1971); Władysław Smolski, *Zaklęte Lata* [Cursed Years] (Warszawa: Pax, 1964); Isaiah Trunk, *Jewish Responses to Nazi Persecutions* (New York: Stein and Day Publishers, 1979). The publications by Boder and Trunk contain only very few accounts about Jews who passed and none about rescuers.

4. To be sure, the Holocaust literature contains many more descriptive accounts of survivors, but such publications rarely focus on Poland or on the rescuers and rescued. See, for example, Azriel Eisenberg, *Witness to the Holocaust* (New York: The Pilgrim Press, 1981); Sylvia Rothchild, ed., *Voices from the Holocaust* (New York: New American Library, 1981).

5. A few of these memoirs were written by people who died during the war, a few by people who died since the war. By including such cases I have tried to diversify the sample.

6. Here and there, especially in secondary sources, there was mention of Polish anti-Semites, who during the war risked their lives to save Jews.

7. For example, when we refer to an apartment that was unsafe for Jews—that is a place known to the Nazis as a Jewish refuge—we refer to it as "burned," a "burned place." Jews who were protected by Poles were called by the latter "cats."

8. To those I interviewed directly I promised anonymity. Therefore, I use fictitious names. Also when I refer to people by name and quote them without giving a footnote, this means that I have interviewed them directly. I intend to donate all of my tapes and transcripts to YIVO, Institute of Jewish Research, New York City, with the stipulation that the real names be revealed only after thirty years. I hope that this will protect the anonymity of the sixty-five individuals I interviewed.

9. Testimonies of these particular individuals were a part of the institute's archives recorded much earlier. I compared these testimonies to my interviews, and found without exception that the information in the archives was more limited, referring to a few facts only. In no case, however, did I find serious distortions, only omissions.

10. Even though those I interviewed came from different walks of life, I do not claim that these groups are representative of the survivors or rescuers. Not all Jews who survived by passing, or Poles who were involved in aiding Jews, cared to identify themselves after the war. According to one estimate, after the war 20,000 Jews never admitted to being Jewish. See Bernard D. Weinryb, "The Jews in Poland," pp. 207–326 in *The Jews in the Soviet Satellite* eds. Peter Meyer, et al. (Westport, Conn.: Greenwood Press, 1953). As for the rescuers, some Poles were afraid of other Poles and kept their saving of Jews a secret. Also, within each group many must have died since the end of the war. For these and other reasons there is simply no way in which I or anyone else could have had a representative sample. The next best thing to representativeness, however, is diversification and this is what I tried to achieve.

11. The following table is self-explanatory:

The rescued and the rescuers

Sources of material	Jews* (%)	Poles (%)
Brief published accounts and case histories	17	39
Memoirs published	13	16
Archives YIVO	26	12
Archives Jewish Historical Institute	17	6
Yad Vashem	16	10
Direct interviews	11	17
	$n = 308$	$n = 189$

*The 308 Jewish survivors discuss 565 Polish helpers.

12. By and large, Poles referred to the Jews they protected in general terms, whereas the Jews were more likely to describe their protectors in personal terms. Because of my special interest in learning about the attributes and motivations of rescuers, whenever possible I carefully recorded such descriptions. For the same reason the Jewish survivors were asked about different characteristics of Polish

helpers. The entire Jewish sample contains descriptions of 565 helpers. These are used as additional and indirect cases of rescuers.

13. To make the data manageable, for each sample I devised a set of coding categories. When developing these coding categories I relied both on the actual data and on the kind of information I would have liked to have had. The specific categories were developed in a number of steps, taking at each step a few random cases which I then reorganized in terms of more general categories. Given the heterogeneity of data, some cases had a high level of "no information." Eventually the coded information was punched into IBM cards. One card was used per individual with seventy-five columns as the absolute maximum.

Selected Bibliography

Articles

Agar, Herbert. "Christians Who Dared Death to Save Jews." *Jewish Digest,* May 1961, pp. 7–11.

Ainsztein, Reuben. "The Bandera-Oberaender Case." *Midstream,* no. 2 (1960): 17–25.

Ainsztein, Reuben. "The Jews of Poland Need Not Have Died." *Midstream,* Autumn 1958, pp. 2–4, 101–103.

Aldridge, James. "Poles Abuse Jews." *Liberal Judaism,* April 1944, pp. 50–52.

Alexander, L. "War Crimes: Their Social Psychological Aspects." *American Journal of Psychiatry* 39, no. 3 (September 1948): 170–177.

Altmann, S.P. "Simmel's Philosophy of Money." *American Journal of Sociology* 9 (1903): 46–68.

Anielewicz, Mordechai. "The Jewish Fighter's Organization Appeals." *Extermination and Resistance, Historical Records and Source Material* 1 (1958): 7–9.

Auni, Haim. "The Zionist Underground in Holland and France and the Escape to Spain." In *Rescue Attempts during the Holocaust,* edited by Yisrael Gutman and Efrâim Zuroff, pp. 555–590. Jerusalem: Yad Vashem, 1977.

Baron, Lawrence. "Restoring Faith in Humankind." *S'HMA* 14 (September 1984): 124–128.

Bauminger, Arieh. "Righteous Gentiles." *Jewish Spectator* 27–31 (September 1964): 9–10.

Bejski, Moshe. "The Righteous among the Nations and Their Part in the Rescue of Jews." In *The Catastrophe of European Jewry,* edited by Yisrael Gutman and Livia Rothkirchen, pp. 582–607. Jerusalem: Yad Vashem, 1976.

Berkowitz, Leonard. "The Self, Selfishness and Altruism." In *Altruism and Helping Behavior,* edited by J.R. Macaulay and L. Berkowitz, pp. 143–151. New York: Academic Press, 1970.

Berkowitz, Leonard. "Social Norms, Feelings and Other Factors Affecting Helping and Altruism." In *Cooperation and Helping Behavior Theories and Research,* edited by V.J. Derlega and J. Grzelak, pp. 63–108. New York: Academic Press, 1982.

Berman, Adolf. "The Fate of the Children in the Warsaw Ghetto." In *The Catastrophe of European Jewry*, edited by Yisrael Gutman and Livia Rothkirchen, pp. 400–421. Jerusalem: Yad Vashem, 1976.

Bettelheim, Bruno. "Freedom from Ghetto Thinking." *Midstream*, Spring 1962, pp. 16–25.

Bettelheim, Bruno. "Individual and Mass Behavior in Extreme Situations." *Journal of Abnormal and Social Psychology* 38 (October 1943): 417–452.

Bickman, Leonard. "Effects of Another Bystander's Ability to Help in Bystander Intervention in an Emergency." *Journal of Experimental Social Psychology* 6 (May 1971): 367–379.

Bickman, Leonard. "Social Influence and Diffusion of Responsibility." *Journal of Experimental Social Psychology* 8, no. 5 (September 1972): 438–445.

Bickman, Leonard, and Kamzan, Mark. "The Effect of Race and Need on Helping Behavior." *The Journal of Social Psychology* 89 (1973): 73–77.

Bitner, Wacław. "Catholics and the Jews in Poland." *Jewish Forum*, August 1944, p. 164.

Blau, Peter M. "Social Exchange." In *International Encyclopedia of the Social Sciences*, edited by David L. Sills, pp. 452–457. New York: The Macmillan Co., 1968.

Bluhm, Hilde O. "How Did They Survive?" *American Journal of Psychiatry* 2, no. 1 (1948): 3–32.

Blumenthal, Nachman. "The Plight of The Jewish Partisans." *Yad Vashem Bulletin*, no. 1 (April 1957): 4–7.

Brenner, Arline R., and Levin, James M. "Off-Duty Policemen and Bystander Apathy." *Journal of Police Science and Administration* 1, no. 1 (March 1973): 61–64.

Bryan, James H., and London, Perry. "Altruistic Behavior by Children." *Psychological Bulletin* 75, no. 3 (1970): 200–211.

Bryan, James H., and Test, M.A. "Models of Helping: Naturalistic Studies in Aiding Behavior." *Journal of Personality and Social Psychology* 6 (1967): 400–407.

"Champion of Justice, Irene Harand." *Wiener Library Bulletin* 9 (1955): 24.

Darley, John, and Latane, Bibb. "When Will People Help?" *Psychology Today* 2, no. 7 (1968): 54, 56–57, 70–71.

Dobroszycki, Lucjan. "Restoring Jewish Life in Post War Poland." *Soviet Jewish Affairs* 3, no. 2 (1973): 1–15.

Dworzecki, Mark. "The Day to Day Stand of the Jews." In *The Catastrophe of European Jewry*, edited by Yisrael Gutman and Livia Rothkirchen, pp. 367–399. Jerusalem: Yad Vashem, 1976.

Ehlert, Jeff; Ehlert, Nuala; and Merrens, Matthew. "The Influence of Ideological Affiliation on Helping Behavior." *The Journal of Social Psychology* 89 (1973): 315–316.

Feigenbaum, M.J. "Life in a Bunker." In *The Root and the Bough*, edited by Leo W. Schwarz, pp. 145–146. New York: Rinehart and Co., Inc., 1949.

Fellner, Carl H., and Marshall, John R. "Kidney Donors." In *Altruism and Helping Behavior*, edited by J.R. Macaulay and L. Berkowitz, pp. 269–281. New York: Academic Press, 1970.

Fink, Carole. "Germany and the Polish Elections of November 1930: A Study in League Diplomacy." *East European Quarterly* 15, no. 2 (June 1981): 181–207.

Fogelman, Eva. "Social Psychological Study of the Rescuers." *Martyrdom and Resistance*, September 1984, p. 8.

Forbes, Gordon B.; Tevault, R.; Kent, Gromell Henry F. "Willingness to Help Strangers as a Function of Liberal Conservative or Catholic Church Membership: A Field Study with the Lost Letter Technique." *Psychological Reports* 28 (1971): 947–949.

Freud, Anna. "A Form of Altruism." In *The Ego And the Mechanism of Defense*, edited by Anna Freud, pp. 132–148. New York: International Universities Press, Inc., 1946.

Friedman, Philip. "Aspects of Research in the Judenrat." In *Roads to Extinction: Essays on the Holocaust*, edited by Ada June Friedman, pp. 539–553. Philadelphia: The Jewish Publications Society of America, 1980.

Friedman, Philip. "The Lubian Reservation and the Madagascar Policy." *YIVO Annual of Social Science* 8 (1958): 151–177.

Friedman, Philip. "Mordechai Chaim Rumkowski of Łódź." In *Roads to Extinction: Essays on the Holocaust*, edited by Ada June Friedman, pp. 353–364. Philadelphia: The Jewish Publications Society of America, 1980.

Friedman, Philip. "Righteous Gentiles in the Nazi Era." In *Roads to Extinction: Essays on the Holocaust*, edited by Ada June Friedman, pp. 409–421. Philadelphia: The Jewish Publications Society of America, 1980.

Fry, Varian. "Operation Emergency Rescue." *The New Leader*, December 20, 1965, pp. 11–14.

Gouldner, Alvin W. "The Norm of Reciprocity: A Preliminary Statement." *American Sociological Review* 25, no. 2 (April 1960): 161–178.

Gross, Nathan. "Aryan Papers in Poland." *Extermination and Resistance, Historical Sources and Material* 1 (1958): 79–86.

Gross, Nathan. "Days and Nights in the 'Aryan' Quarter: The Daily Worries of a Jew Carrying Aryan Papers." *Yad Vashem Bulletin*, no. 4/5 (1959): 12–13.

Gross, Nathan. "Research into the Question of Aryan Papers." *Yediot Yad Vashem*, no. 10–11 (1956): 34.

Gross, Nathan. "Unlucky Clara." *Yad Vashem Bulletin*, no. 15 (August 1964): 55–60.

Grossman, Kurt R. "The Humanitarian That Cheated Hitler." *Coronet*, September 1959, pp. 66–71.

Grossman, Kurt R. "What Were the Jewish Losses?" *Congress Weekly*, October 12, 1953, pp. 9–11.

Gruber, Ruth M. "The Pole Who Saved 32 Jews." *Hadassah Magazine*, December 1968, pp. 21, 36.

Hackler, James C.; Ho Kwai-Yiu; and Urquhart-Ross, Carol. "The Willingness to Intervene." *Social Problems* 21, no. 3, (1974): 328–344.

Harris, Mary B. "The Effects of Performing One Altruistic Act on the Likelihood of Performing Another." *Journal of Social Psychology* 88 (1972): 65–73.

Heller, Celia S. "Poles of Jewish Background—The Case of Assimilation without Integration in Prewar Poland." In *Studies on Polish Jewry*, edited by J.A. Fishman, pp. 242–276. New York: YIVO Institute for Jewish Research, 1974.

Tadeusz Hołuj, "Jan Mosdorf's Truce." In *Lest We Forget*, edited by Adolf Rudnicki, pp. 118–124. Warsaw: Polonia Foreign Languages Publishing House, 1955.

Huneke, Douglas. "A Study of Christians Who Rescued Jews during the Nazi Era." *Humboldt Journal of Social Relations* 9, no. 1 (1981–1982): 144–150.

Huston, Ted L.; Geis, Gilbert; and Wright, Richard. "The Angry Samaritans." *Psychology Today*, June 1976, pp. 61–64, 81.

"In the Face of a Common Foe, Poles and Jews during the Occupation." *Wiener Library Bulletin* 15, no. 1 (1961): 10.

"Is Polish Anti-Semitism Incurable?" *Congress Weekly*, December 3, 1943, pp. 50–52.

Jong, Louis De. "The Netherlands and Auschwitz: Why Were the Reports of Mass Killings so Widely Disbelieved?" In *Imposed Jewish Governing Bodies under Nazi Rule*, pp. 11–30. YIVO Colloquium, December 1967.

Kanabus, Felix. "Address of the JNF September 20, 1964." In *Anthology of Holocaust Literature*, edited by Jacob Gladstein, et al., pp. 392–395. Philadelphia: The Jewish Publications Society of America, 1969.

Kanner, Feiga. "Winter in The Forest." In *Anthology of Holocaust Literature*, edited by J. Gladstein, et al., pp. 155–157. Philadelphia: The Jewish Publications Society of America, 1969.

Karlykowski, Jerzy. "Two Types of Altruistic Behavior: Doing Good to Feel Good or to Make the Other Feel Good." In *Cooperation and Helping Behavior Theories and Research*, edited by V.J. Derlega and J. Grzelak, pp. 397–413. New York: Academic Press, 1982.

Kermish, Joseph. "The Activities of the Council for Aid to Jews (Żegota) in Occupied Poland." In *Rescue Attempts during The Holocaust*, edited by Yisrael Gutman and Efraim Zuroff, pp. 367–398. Jerusalem: Yad Vashem, 1977.

Krebs, Dennis L. "Altruism—An Examination of the Concept and a Review of the Literature." *Psychological Bulletin* 73, 1970: 258–302.

Korzec, Paweł. "Anti-Semitism in Poland as an Intellectual, Social and Political Movement." In *Studies on Polish Jewry*, edited by Joshua A. Fishman, pp. 12–114. New York: YIVO Institute for Jewish Research, 1974.

Latane, Bibb, and Rodin, Judith. "A Lady in Distress: Inhibiting Effects of Friends and Strangers on Bystander Intervention." *Journal of Experimental Social Psychology* 5 (1969): 189–202.

Lazarsfeld, Paul F., and Merton, Robert K. "Friendship as Social Process: A Substantive and Methodological Analysis." In *Freedom and Control in Modern Society*, edited by Berger, et al., pp. 18–66. New York: Van Nostrand, 1954.

Lędnicki, Wacław. "Poles and Jews before and after the War." *The Jewish Forum*, August 1944, pp. 151–153.

Leeds, Ruth. "Altruism and the Norm of Giving." *Merrill-Palmer Quarterly* 9 (1963): 229–240.

Leonard, Oscar. "Anti-Semitism Disgraces Christianity." *National Jewish Monthly*, February 1937, p. 157.

Lestchinsky, Jacob. "Balance Sheet of Extermination." *Jewish Affairs* 1, no. 1 (February 1946): 10.

Lestchinsky, Jacob. "Economic Aspects of Jewish Community Organization in Independent Poland." *Jewish Social Studies* 9, no. 1–4 (1947): 319–338.

Lestchinsky, Jacob. "The Industrial and Social Structure of the Jewish Population of Interbellum Poland." *YIVO Annual of Jewish Social Science* 2 (1956–1957): 243–269.

Lestchinsky, Jacob. "The Jews in the City of the Republic of Poland." *YIVO Annual of Jewish Social Science* 1 (1946): 156–177.

Lewin, Kurt. "Self-Hatred among Jews." *Contemporary Jewish Record* 4 (June 1941): 219–232.

Lichten, Joseph L. "Did Polish Jews Die Forsaken?" *The Polish Review* 4, no. 1–2 (1959): 119–126.

Lifschutz, L. "Selected Documents Pertaining to Jewish Life in Poland, 1919–1938." In *Studies on Polish Jewry*, edited by Joshua A. Fishman, pp. 277–294. New York: YIVO Institute for Jewish Research, 1974.

Lipser, Abraham. "Escape." In *The Root and the Bough*, edited by Leo W. Schwarz, pp. 304–307. New York: Holt, Rinehardt and Co., 1949.

London, Perry. "The Rescuers: Motivational Hypotheses about Christians Who Saved Jews from the Nazis." In *Altruism and Helping Behavior*, edited by J.R. Macaulay and L. Berkowitz, pp. 241–250. New York: Academic Press, 1970.

Lowe, Roland, and Richter, Gary. "Relation of Altruism to Age, Social Class and Ethnic Identity." *Psychological Reports* 33 (1973): 567–572.

Macaulay, Jacqueline R., and Berkowitz, Leonard. "Overview." In *Altruism and Helping Behavior*, edited by J.R. Macaulay and L. Berkowitz, pp. 1–9. New York: Academic Press, 1970.

Mahler, Ella. "How Granny Saved Helenka from the Germans." In *Flame and Fury*, edited by Yaakov Shilhav and Sara Feinstein, pp. 77–82. New York: Jewish Education Committee Press, 1962.

Mason, Henry L. "Jews in the Occupied Netherlands." *Political Science Quarterly* 99, no. 2 (Summer 1984): 315–343.

Midlarski, Elizabeth. "Aiding Responses: An Analysis and Review." *Merrill Palmer Quarterly* 14, (1968): 229–260.

Oliner, Samuel P. "The Heroes of the Nazi Era: A Plea for Recognition." *Reconstructionist* 48 (June 1982): 7–14.

Oliner, Samuel P. "The Unsung Heroes in Nazi-Occupied Europe: The Antidote for Evil." *Nationalities Papers* 12, no. 1 (Spring 1984): 129–136.

Piliavin, Jane Allyn; Dovidio, John F.; Gaetner, Samuel L.; and Clark, Russel D. III. "Responsive Bystanders: The Process of Intervention." In *Cooperation and Helping Behavior Theories and Research*, edited by V.J. Derlega and J. Grzelak, pp. 279–304. New York: Academic Press, 1982.

Pinczewska, Rosa. "Rosa's Journey." In *The Root and the Bough*, edited by Leo W. Schwarz, pp. 297–299. New York: Rinehart and Co., Inc., 1949.

Poliakov, Leon. "Human Morality and the Nazi Terror." *Commentary*, August 1950, pp. 111–116.

Poliakov, Leon. "The Vatican and the Jewish Question." *Commentary*, November 1950, pp. 439–449.

"Polish Underground Liquidates Betrayers of Jews." *Ghetto Speaks*, no. 27 (July 15, 1944): 3.

Pollack, S.W. "Need They Have Died?" *World Jewry* (November 1958): 14–15.

Rosenhan, David L. "The Natural Socialization of Altruistic Autonomy." In *Altruism and Helping Behavior*, edited by J.R. Macaulay and L. Berkowitz, pp. 251–268. New York: Academic Press, 1970.

Rosenhan, David L. "Towards Resolving the Altruism Paradox: Affect, Self-Reinforcement and Cognition." In *Sympathy and Helping: Psychological and Social Principles*, edited by Lauren Wispe, pp. 101–113. New York: Academic Press, 1978.

Rothkirchen, Livia. "Czech Attitudes towards the Jews during Nazi Regime," *Yad Vashem Studies* 13 (1979): 287–320.

Rubinstein, Sonia. "Porick." In *The Root and the Bough*, edited by Leo W. Schwarz, pp. 180–183. New York: Rinehart and Co., Inc., 1949.

Sapetowa, Karla. "A Polish Woman Relates Her Story." In *Anthology of Holocaust Literature*, edited by J. Gladstein, et al., pp. 370–372. Philadelphia: The Jewish Publications Society of America, 1969.

Schachter, Stanley J. "Bettelheim and Frankel: Contradicting Views of the Holocaust." *Reconstructionist* 26, no. 20 (February 10, 1961): 6–11.

Schneider, Frank W., and Mokus, Zig. "Failure to Find a Rural Urban Difference in Incidence of Altruistic Behavior." *Psychological Reports* 35 (1974): 294.

Sijes, B.A. "Several Observations Concerning the Position of the Jews in Occupied Holland during World War II." In *Rescue Attempts during the Holocaust*, edited by Yisrael Gutman and Efraim Zuroff, pp. 527–553. Jerusalem: Yad Vashem, 1977.

Sorokin, Pitrim A. "Factors of Altruism and Egoism." *Social Science Research* 32 (1948): 674–678.

Sorokin, Pitrim, A. "Similarity and Dissimilarity as Factors of Altruism." *Social Science Research* 32 (1948): 776–781.

Sutzkever, Arieh. "Never Say This is the Last Road." In *The Root and the Bough*, edited by Leo W. Schwarz, pp. 66–92. New York: Holt, Rinehart and Co., Inc., 1949.

Tec, Nechama. "Righteous Christians in Poland." *International Social Science Review*, Winter 1983, pp. 12–19.

Tec, Nechama. "Righteous Christians—Who Are They?" In *Book of Proceedings, Eighth World Congress of Jewish Studies*. Jerusalem, 1982, pp. 167–172.

Tec, Nechama. "Sex Distinctions and Passing as Christians during the Holocaust." *East European Quarterly* 18, no. 1 (March 1984): 113–123.

Tec, Nechama. "Polish Anti-Semitism and Christian Protectors." *East European Quarterly*, forthcoming.

"The JFO and the Polish Underground." *Extermination and Resistance, Historical Records and Source Material* 1 (1958): 14–19.

Valentin, Hugo. "Rescue and Relief Activities in Behalf of Jewish Victims of Nazism in Scandinavia." *YIVO Annual of Social Science* 8 (1953): 224–251.

Weigler, Zvi. "Two Villages Razed for Extending Help to Jews." *Yad Vashem Bulletin*, no. 1 (1957): 18–20.

Weinryb, Bernard D. "The Jews in Poland." In *The Jews in the Soviet Satellite*, edited by Peter Meyer et al., pp. 207–326. Westport, Conn.: Greenwood Press, 1953.

Wilson, Edward O. "The Genetic Evolution of Altruism." In *Sympathy and Helping: Psychological and Sociological Principles*, edited by Lauren Wispe, pp. 11–37. New York: Academic Press, 1978.

Wispe, Lauren. "Toward an Integration." In *Sympathy and Helping: Psychological and Sociological Principles*, edited by Lauren Wispe, pp. 303–327. New York: Academic Press, 1978.

Wynot, Edward D., Jr. "A Necessary Cruelty: The Emergence of Official Anti-Semitism in Poland, 1936–1939." *The American Historical Review* 76, no. 4 (October 1971): 1035–1058.

Books

Abel, Theodore. *The Nazi Movement*. New York: Atherton Press, 1966.

Adler, Stanisław. *In the Warsaw Ghetto*. Jerusalem: Yad Vashem, 1982.

Ainsztein, Reuben. *Jewish Resistance in Nazi-Occupied Eastern Europe*. New York: Barnes and Noble, 1974.

Aizenberg, Lisa. *One in a Million*. Berkeley, Calif.: Judah L. Magnes Memorial Museum, 1968.

Anger, Per. *With Raoul Wallenberg in Budapest*. New York: Holocaust Library, 1981.

Arad, Yizhak. *Ghetto in Flames: The Struggle and Destruction of the Jews in Vilna in the Holocaust*. New York: Holocaust Library, 1982.

Arad, Yizhak; Gutman, Yisrael; and Margaliot, Abraham, eds. *Documents of the Holocaust*. Jerusalem: Yad Vashem, 1981.

Arendt, Hannah. *Eichmann in Jerusalem: A Report on the Banality of Evil*. New York: Penguin Books, 1978.

Arendt, Hannah. *The Jew as Pariah*. New York: Grove Press, Inc., 1978.

Arendt, Hannah. *The Origins of Totalitarianism*. New York: Harcourt Brace Jovanovich, 1973.

Arieti, Silvano. *The Parnas*. New York: Basic Books, Inc., 1979.

Aycoberry, Pierre. *The Nazi Question*. New York: Pantheon Books, 1981.

Bar Oni, Bryna. *The Vapor*. Chicago: Visual Impact, Inc., 1976.

Bartoszewski, Władysław. *The Blood Shed Unites Us*. Warsaw: Interpress Publishers, 1970.

Bartoszewski, Władysław, and Lewinówna, Zofia. *The Samaritans: Heroes of the Holocaust*. New York: Twyne Publishers, Inc., 1970.

Bauer, Yehuda. *The Holocaust in Historical Perspective*. Seattle: University of Washington Press, 1978.

Bauminger, Arieh L. *Roll of Honor*. Jerusalem: Yad Vashem, 1970.

Berenstein, Tatiana, and Rutkowski, Adam. *Assistance to the Jews in Poland, 1939–1945*. Warsaw: Polonia Foreign Languages Publishing House, 1963.

Berg, Mary. *Warsaw Ghetto: A Diary*. New York: L.B. Fischer, 1945.

Berger, Zdena. *Tell Me Another Morning*. New York: Harper and Row, 1961.

Bettelheim, Bruno. *The Informed Heart*. Glencoe, Ill.: The Free Press, 1960.

Bierman, John. *Righteous Gentile: The Story of Raoul Wallenberg, Missing Hero of the Holocaust*. New York: The Viking Press, 1981.

Boder, David P. *I Did Not Interview the Dead*. Urbana, Ill.: University of Illinois Press, 1949.

Boem, Erich, ed. *We Survived*. Santa Barbara, Calif.: Clio, 1966.

Boom, Corrie ten, with John and Elizabeth Sherill. *The Hiding Place*. Old Tappan, N.J.: Fleming H. Revell Co. 1971.

Borzykowski, Tuvia. *Between Falling Walls*. Israel: Ghetto Fighters' House, 1972.

Brand, Sandra. *I Dared to Live*. New York: Shengold Publishers, Inc., 1978.

Cargas, Harry James. *A Christian Response to the Holocaust*. Denver, Colo.: Stonehenge Books, 1981.

Cargas, Harry James, ed. *When God and Man Failed*. New York: The Macmillan Co., 1981.

Chciuk, Andrzej, ed. *Saving Jews in Wartime Poland*. Melbourne, Australia: Wilkie Co. Ltd., 1969.

Cohen, Elie A. *Human Behavior in the Concentration Camp*. London: Johnathan Cape, 1954.

David, Janina. *A Touch of Earth: A Wartime Childhood*. New York: Orion Press, 1969.

Dawidowicz, Lucy, ed. *A Holocaust Reader*. New York: Behrman House, Inc., 1976.

Dawidowicz, Lucy. *The Jewish Presence: Essays on Identity and History*. New York: Holt, Rinehart and Winston, 1977.

Dawidowicz, Lucy S. *The War against the Jews 1939–1945*. New York: Holt, Rinehart and Winston, 1975.

Davies, Norman. *Heart of Europe: A Short History of Poland*. Oxford: Clarendon Press, 1984.

Derlega, Valerian J., and Grzelak, Janusz, eds. *Cooperation and Helping, Behavior Theories and Research*. New York: Academic Press, 1982.

Feingold, Henry L. *The Politics of Rescue*. New York: Holocaust Library, 1970.

Fenelon, Fania. *Playing for Time*. New York: Atheneum Publishers, 1977.

Fishman, Joshua A., ed. *Studies on Polish Jewry*. New York: YIVO Institute for Jewish Research, 1974.

Fleming, Gerald. *Hitler and the Final Solution*. Berkeley, Calif.: University of California Press, 1984.

Flender, Harold. *Rescue in Denmark*. New York: Simon and Schuster, 1963.

Flinker, Moshe. *Moshe's Diary*. Jerusalem: Yad Vashem, 1971.

Frank, Anne. *The Diary of a Young Girl*. New York: Washington Square Press, 1972.

Frankl, Victor E. *Man's Search for Meaning*. New York: Simon and Schuster, 1959.

Friedlander, Saul. *When Memory Comes*. New York: Farrar, Straus and Giroux, 1979.

Friedlander, Saul. *Pius XII and the Third Reich: A Documentation*. New York: Octagon, 1980.

Friedman, Ada June, ed. *Roads to Extinction: Essays on the Holocaust*. Philadelphia: The Jewish Publications Society of America, 1980.

Friedman, Ina R. *Escape or Die: True Stories of Young People Who Survived the Holocaust*. Reading, Mass.: Addison-Wesley, 1982.

Friedman, Philip. *Their Brothers' Keepers*. New York: Holocaust Library, 1978.

Des Pres, Terrence. *The Survivor: The Anatomy of Life in the Death Camps*. New York: Oxford University Press, 1976.

Dobroszycki, Lucjan, ed. *The Chronicle of the Lódź Ghetto, 1941–1944*. New Haven, Conn.: Yale University Press, 1984.

Donat, Alexander, (ed. *The Death Camp Treblinka*. New York: Holocaust Library, 1979.

Donat, Alexander. *The Holocaust Kingdom: A Memoir*. New York: Holt, Rinehart and Winston, 1965.

Durkheim, Emile. *The Rules of Sociological Method*. Glencoe, Ill.: The Free Press, 1938.

Durkheim, Emile. *Suicide*. Glencoe, Ill.: The Free Press, 1951.

Eckhart, Arthur Roy. *Your People, My People*. New York: Quadrangle Books, 1974.

Ehrenburg, Ilya, and Grossman, Vasily, eds. *The Black Book: Documents the Nazis' Destruction of 1.5 Million Soviet Jews, Suppressed by Stalin*. New York: Holocaust Library, 1981.

Eisenberg, Azriel. *Witness to the Holocaust*. New York: The Pilgrim Press, 1981.

Engelman, Bernt. *Germany without Jews*. New York: Bantam Books, 1984.

Epstein, Helen. *Children of the Holocaust*. New York: G.P. Putnam and Sons, 1979.

Fein, Helen. *Accounting for Genocide*. New York: The Free Press, 1979.

Fuks, Marian; Hoffman, Zygmunt; Horn, Maurycy; and Tomaszewski, Jerzy. *Polish Jewry, History and Culture*. Warsaw: Interpress Publishers, 1982.

Gershon, Karen, ed. *We Came as Children: A Collective Autobiography*. London: Gollanch, 1966.

Gerson, Louis L. *Woodrow Wilson and the Rebirth of Poland, 1914–1920*. Hamden, Conn.: Archon Books, 1972.

Gilbert, Martin. *Auschwitz and the Allies: A Devastating Account of How the Allies Responded to the News of Hitler's Mass Murder*. New York: Holt, Rinehart and Winston, 1981.

Gladstein, Jacob; Knox, Israel; and Margoshes, Samuel. *Anthology of Holocaust Literature*. Philadelphia: The Jewish Publications Society of America, 1969.

Glück, Gemma Laguardia. *My Story*. New York: David McKay, 1961.

Goldstein, Bernard. *The Stars Bear Witness*. New York: The Viking Press, 1949.

Gray, Martin, and Gallo, Max. *For Those I Loved*. Boston: Little, Brown and Co., 1972.

Green, Gerald. *The Artists of Terezin*. New York: Schocken Books, 1978.

Gross, Leonard. *The Last Jews in Berlin*. New York: Simon and Schuster, 1982.

Grynberg, Henryk. *Childhood of Shadows*. London: Vallentine, Mitchell, 1969.

Gurdus-Krugman, Luba. *The Death Train*. New York: Holocaust Library, 1978.

Gutman, Yisrael, and Rothkirchen, Livia, eds. *The Catastrophe of European Jewry*. Jerusalem: Yad Vashem, 1976.

Gutman, Yisrael, and Zuroff, Efraim, eds. *Rescue Attempts during the Holocaust*. Jerusalem: Yad Vashem, 1977.

Hallie, Phillip. *Lest Innocent Blood Be Shed*. New York: Harper and Row, 1979.

Hardin, Garrett. *The Limits of Altruism: An Ecologist's View of Survival*. Bloomington, Ind.: Indiana University Press, 1977.

Hausner, Gideon. *Justice in Jerusalem*. New York: Holocaust Library, 1966.

Heller, Celia S. *On the Edge of Destruction*. New York: Columbia University Press, 1977.

Hellman, Peter. *Avenue of the Righteous*. New York: Atheneum Publishers, 1980.

Hilberg, Raul. *The Destruction of the European Jews*. New York: New Viewpoints, 1973.

Hilberg, Raul, ed. *Documents of Destruction*. New York: Quadrangle Books, 1971.

Hilberg, Raul; Staron, Stanislaw; and Kermisz, Jósef, eds. *The Warsaw Diary of Adam Czerniakow*. New York: Stein and Day Publishers, 1979.

Hillesum, Etty. *An Interrupted Life: The Diaries of Etty Hillesum, 1941–1943*. New York: Pantheon Books, 1983.

Hochhuth, Rolf. *The Deputy*. New York: Grove Press, 1964.

Horbach, Michael. *Out Of the Night*. New York: Frederick Fell, Inc., 1967.

Iranek-Osmecki, Kazimierz. *He Who Saves One Life*. New York: Crown Publishers, Inc., 1971.

Iwańska, Alicja. *Contemporary Poland, Society, Politics, Economy*. Chicago: The Chicago University Press, 1955.

Jackson-Bitton, Livia E. *Elli: Coming of Age in the Holocaust*. New York: Quadrangle Books, 1980.

Jacobs, Joseph. *Jewish Contributions to Civilization*. Philadelphia: The Jewish Publications Society of America, 1919.

Jaffre, Jeffrey. *Stepson of a People*. Montreal: Admiral Printing, 1968.

Ka-Tzetnik 135638. *Sunrise over Hell*. London: A. Howard and Wyndman Co., 1977.

Kaplan, Chaim. *Scrolls of Agony: The Warsaw Diary of Chaim Kaplan*, ed. and trans. by Abram Katsh. New York: The Macmillan Co., 1956.

Karski, Jan. *Story of a Secret State*. Boston: Houghton Mifflin Co., 1944.

Katzenelson, Yitzhak. *Vittel Diary*. Israel: Ghetto Fighters' House, 1972.

Keneally, Thomas. *Schindler's List*. New York: Simon and Schuster, 1982.

Kersten, Felix, and Briffault, Herman. *The Memoirs of Dr. Felix Kersten*. New York: Doubleday, 1947.

Kielar, Wiesław. *Anus Mundi: 1500 Days in Auschwitz/Birkenau*. New York: Times Books, 1980.

Klein, Catherine. *Escape from Berlin*. London: Victor Gollancz, Ltd., 1944.

Klein-Weissman, Gerda. *All but My Life*. New York: Hill and Wang, 1975.

Kogon, Eugen. *The Theory and Practice of Hell*. Berkeley, Calif.: Berkeley Publishing Corp., 1980.

Kolnicki, Aryeh. *The Diary of Adam's Father*. Israel: Ghetto Fighters' House, 1973.

Komarowski-Bor, Tadeusz. *The Secret Army*. London: Victor Gollancz, Ltd., 1950.

Korboński, Stefan. *The Polish Underground State: A Guide to the Underground, 1939–1945*. Boulder, Colo.: East European Quarterly, 1978.

Krakowski, Shmuel, and Gutman, Yisrael. *Jews and Poles during World War II* (tentative title). New York: Holocaust Library, forthcoming.

Kuchler-Silverman, L. *One Hundred Children*. New York: Doubleday, 1961.

Kulski, Julian Eugeniusz. *Dying We Live*. New York: Holt, Rinehart and Winston, 1980.

Kuper, Jack. *Child of the Holocaust*. New York: Doubleday, 1968.

Kurzman, Dan. *The Bravest Battle: The Twenty-Eight Days of the Warsaw Ghetto Uprising*. New York: G.P. Putnam and Sons, 1976.

Langer, Lawrence L. *Versions of Survival: The Holocaust and the Human Spirit*. Albany: The State University of New York Press, 1982.

Lapide, Pinchas. *Three Popes and the Jews*. New York: Hawthorne Books, Inc., 1967.

Laqueur, Walter. *The Terrible Secret: Suppression of the Truth about Hitler's Final Solution*. Boston: Little, Brown and Co., 1980.

Leitner, Isabella. *Fragments of Isabella*. New York: Thomas and Crowell Publishers, 1978.

Lendvai, Paul. *Anti-Semitism without Jews*. New York: Doubleday, 1971.

Lestchinsky, Jacob. *Crisis, Catastrophe and Survival: A Jewish Balance Sheet, 1914–1948*. New York: Institute of Jewish Affairs of the World Jewish Congress, 1948.

Leuner, H.D. *When Compassion Was a Crime*. London: Oswald Wolf, 1966.

Levin, Nora. *The Holocaust*. New York: Schocken Books, 1973.

Lind, Jakov. *Counting My Steps*. London: Panther Books, Ltd., 1972.

Littell, Franklin H., and Locke, Hubert G., eds. *The German Struggle and the Holocaust*. Detroit, Mich.: Wayne State University Press, 1974.

Macaulay, Jacqueline R., and Berkowitz, Leonard, eds. *Altruism and Helping Behavior*. New York: Academic Press, 1970.

Marrus, Michael R., and Paxton, Robert O. *Vichy France and the Jews*. New York: Schocken Books, 1983.

Meed, Vladka. *On Both Sides of the Wall*. New York: Ghetto Fighters' House and Hakibbutz Hameuhad Publishing House, 1972.

Mendelson, Ezra. *The Jews of East Central Europe: Between the World Wars*. Bloomington, Ind.: Indiana University Press, 1983.

Meyer, Peter; Weinryb, Bernard D.; Duszynski, Eugene; and Sullivan, Nicolas, eds. *The Jews in the Soviet Satellite*. Westport, Conn.: Greenwood Press, 1953.

Minco, Marga. *Bitter Herbs*. New York: Oxford University Press, 1960.

Mizruchi, Ephraim H. *Regulating Society: Marginality and Social Control in Historical Perspective*. New York: The Free Press, 1983.

Morley, John F. *Vatican Diplomacy and the Jews during the Holocaust, 1939–1943*. New York: KTAV Publishing House, Inc., 1980.

Morse, Arthur D. *Why Six Million Died: A Chronicle of American Apathy*. New York: Hart Publishing Co., Inc., 1967.

Mosse, George L. *Toward the Final Solution: A History of European Racism*. New York: Howard Fertig, 1978.

Novitch, Miriam. *Sobibor*. New York: Holocaust Library, 1980.

Nowak, Jan. *Courier from Warsaw*. Detroit, Mich.: Wayne State University Press, 1982.

Nyiszli, Miklos. *Auschwitz: A Doctor's Eyewitness Account*. New York: Fawcett Crest, 1961.

Pat, Jacob. *Ashes and Fire*. New York: International Universities Press 1947.

Pawłowicz, Sala, and Klose, Kevin. *I Will Survive*. New York: Norton and Co. Inc., 1962.

Pinkus, Oskar. *The House of Ashes*. New York: The World Publishing House Co., 1964.

Pisar, Samuel. *Of Blood and Hope*. Boston: Little, Brown and Co., 1979.

Poliakov, Leon. *Harvest of Hate: The Nazi Program for the Destruction of the Jews of Europe*. New York: Holocaust Library, 1979.

Poliakov, Leon. *The History of Anti-Semitism*. New York: Schocken Books, 1976.

Polonsky, Antony. *Politics in Independent Poland, 1921–1939*. Oxford: The Clarendon Press, 1972.

Presser, Jacob. *The Destruction of the Dutch Jews*. New York: E.P. Dutton and Co., Inc., 1969.

Rabinowitz, Dorothy. *Survivors of the Holocaust Living in America*. New York: Alfred Knopf, 1976.

Ramati, Alexander. *The Assisi Underground: The Priests Who Rescued Jews*. New York: Stein and Day Publishers, 1978.

Rashke, Richard. *Escape from Sobibor: The Heroic Story of the Jews Who Escaped from a Nazi Death Camp*. Boston: Houghton Mifflin Co., 1982.

Reuther-Radford, Rosemary. *Faith and Fratricide: The Theological Roots of Anti-Semitism*. New York: The Seabury Press, 1979.

Ringelblum, Emmanuel. *Notes from the Warsaw Ghetto*. New York: Schocken Books, 1975.

Ringelblum, Emmanuel. *Polish–Jewish Relations during the Second World War*, edited by Józet Kermisz and Shmuel Krakowski. Jerusalem: Yad Vashem, 1974.

Rings, Werner. *Life with the Enemy*. New York: Doubleday, 1982.

Roiter, Howard. *Voices from the Holocaust*. Givataim, Israel: Asurno Press, 1975.

Rose, Leesha. *The Tulips Are Red*. New York: A.S. Barnes and Co., 1978.

Rosen, Donia. *The Forest My Friend*. New York: World Federation of Bergen-Belsen Association, 1971.

Rothchild, Sylvia, ed. *Voices from the Holocaust*. New York: New American Library, 1981.

Rudashevsky, Yitskhok. *The Diary of the Vilna Ghetto, June 1941–April 1943*. Israel: Ghetto Fighters' House, 1973.

Rushton, Phillippe J. *Altruism, Socialization and Society*. Englewood Cliffs, N.J.: Prentice-Hall, Inc., 1980.

Sartre, Jean-Paul. *Anti-Semite and Jew*. New York: Schocken Books, 1948.

Schwarz, Leo W., ed. *The Root and the Bough*. New York: Rinehart and Co., Inc., 1949.

Senger, Valentin. *No. 12 Kaiserhof Strasse*. New York: E.P. Dutton, 1980.

Shilhav, Yaakov, and Feinstein, Sara, eds. *Flame and Fury*. New York: Jewish Education Committee Press, 1962.

Simmel, Georg. *The Philosophy of Money*. Boston: Routledge and Kegan Paul, 1978.

Sorokin, Pitrim A. *Forms and Techniques of Altruistic and Spiritual Growth*. Boston: Beacon Press, 1954.

Steinberg. Lucien. *Not as a Lamb*. Farnborough, Hants, England: Saxon House, D.C. Heath, Ltd., 1970.

Tal, Uriel. *Religious and Anti-Religious Roots of Anti-Semitism*. New York: Leo Beck Institute, 1971.

Tec, Nechama. *Dry Tears: The Story of a Lost Childhood*. New York: Oxford University Press, 1984.

Tenenbaum, Joseph. *In Search of a Lost People*. New York: The Beechhurt Press, 1948.

Titmuss, Richard M. *The Gift Relationship: From Human Blood to Social Policy*. New York: Pantheon Books, 1971.

Tokayer, Marvin, and Swartz, Mary. *The Fugu Plan*. New York: Paddington Press, Ltd., 1979.

Trunk, Isaiah. *Jewish Responses to Nazi Persecutions*. New York: Stein and Day Publishers, 1979.

Trunk, Isaiah. *Judenrat*. New York: The Macmillan Co., 1972.

Vogt, Hannah. *The Burden of Guilt: A Short History of Germany 1914–1945*. New York: Oxford University Press, 1964.

Vrba, Rudolf, and Bestie, Alan. *I Cannot Forgive*. New York: Grove Press, 1964.

Wasserstein, Bernard. *Britain and the Jews of Europe, 1939–1945*. London: Institute for Jewish Affairs, 1979.

Wells, Leon. *The Janowska Road*. New York: The Macmillan Co., 1963.

Wispe, Lauren. *Altruism, Sympathy and Helping: Psychological and Sociological Principles*. New York: Academic Press, 1978.

Wolff, Kurt H., ed. *The Sociology of Georg Simmel*. Glencoe, Ill.: The Free Press, 1950.

Wyman, David S. *The Abandonment of the Jews: America and the Holocaust, 1941–1945*. New York: Pantheon Books, 1984.

Wynot, Edward D., Jr. *Polish Politics in Transition*. Athens, Ga.: University of Georgia Press, 1974.

Wynot, Edward D., Jr. *Warsaw between the World Wars: Profile of the Capital City in a Developing Land*. New York: Columbia University Press, 1983.

Yahil, Leni. *The Rescue of Danish Jewry: Test of Democracy*. Philadelphia: The Jewish Publication Society of America, 1969.

Zawodny, Janusz Z. *Nothing but Honor: The Story of the Warsaw Uprising, 1944*. Stanford, Calif.: Hoover Institution Press, 1979.

Ziemian, Joseph. *The Cigarette Sellers of Three Cross Square*. Minneapolis, Minn.: Lerner Publications Co., 1975.

Zylberberg, Michael. *A Warsaw Diary, 1939–1945*. London: Vallentine, Mitchell, 1969.

Foreign Articles

Andrzejewski, Jerzy. "Martwa Fala" [Dead wave]. In *Dead Wave,* edited by J. Andrzejewski, pp. 20–50. Warszawa: Spółdzielnia Wydawynicza, 1947.

Arczyński, Marek. "Rada Pomocy Żydom W Polsce" [Council for aid to Jews]. *Biuletyn Żydowskiego Instytutu Historycznego* (hereafter referred to as *ZIH*), no. 65–66 (1968): 173–185.

Bartoszewski, Władysław. "*Egzekucje Publiczne W Warszawie W Latach 1943–1944*" [Public executions in Warsaw]. *Biuletyn Głównej Komisji Badania Zbrodni Niemieckiej W Polsce,* no. 6 (1946): 211–224.

Berengaut, Alfred. "O Bohatestwie Eliasza Kupidłowskiego" [About the heroism of Eliasz Kupidłowski]. *ZIH,* no. 85 (1973): 121–126.

Berenstein, Tatiana, and Rutkowski, Adam. "O Ratowaniu Żydow Przez Polaków W Okresie Okupacji Hitlerowskiej" [Polish rescue of Jews during the occupation]. *ZIH,* no. 35, (1960): 3–46.

Bermanowa, Basia. "Pierwsza Irena," Wspomienia O Irenie Sawickiej [Recollections about Irena Sawicka]. *Przełom* 15, no. 10 (1947): 10.

Chodziński, Edward. "Pomoc Żydom Udzielana Przez Konspiracyjne Biuro Fałszywych Dokumentów W Okresie Okupacji Hitlerowskiej" [Help offered to Jews during the war by the underground office of illegal documents]. *ZIH,* no. 75 (1970): 129–132.

Czarnowski, Tadeusz. "Pomoc Ludnosci Żydowskiej Przez Pracowników Wydziału Ludności Zarządu M. St. Warszawy W Okresie Okupacji, 1939–1945" [Help offered to Jews by employees of the Warsaw population Department during the occupation]. *ZIH,* no. 75 (1970): 119–128.

Datner, Szymon. "Materiały Z Dziedziny Ratownictwa Żydow W Polsce W Okresie Okupacji Hitlerowskiej" [Materials about Jewish rescue during the occupation]. *ZIH* 73 (1970): 133–138.

Datner, Szymon. "Relacja O Pomocy Żydom W Czasie Okupacji Hitlerowskiej" [Help extended to Jews during the occupation]. *ZIH,* no. 71–72 (1969): 229–236.

Dumin-Wąsowicz, Janina. "Wspomnienia O Akcji Pomocy Żydom Podczas Okupacji Hitlerowskiej" [Recalling aid to Jews during occupation]. *ZIH,* no. 45–46 (1963): 248–261.

Eisenbach, Artur. "Tajne Pertaktacje Anglo-Amerykańskie Z Faszystowskimi Niemcami O Los Ludnosci Żydowskiej Pod Okupacją Hitlerowska" [Secret Anglo-American negotiations with the Nazis about the fate of the Jews]. *ZIH,* no. 6–7 (1953): 197–217.

Friedman, Filip. "Zagłada Żydow Polskich W Latach 1939–1945" [Destruction of the Polish Jewry, 1939–1945]. *Biuletyn Głównej Komisji Badania Zbrodni Niemieckiej W Polsce,* no. 6 (1946): 165–208.

Fulkowski, Stefan. "Aby Nie Odezwało Się Już Ani Jedno Echo Ludzkiego Barbarzyństwa" [There should be no echoes of human barbarity]. In *Dead Wave,* edited by J. Adrzejewski, pp. 65–70. Warszawa: Spółdzielnia Wydawnicza, 1947.

Goldkorn, Józef. "Sprawiedliwy Wśród Narodów" [The righteous among the nations]. *Nasz Głos,* no. 15 (April 19, 1967): 1, 4.

Górecki, Juliusz. "Męczennikom Gheta" [To ghetto martyrs]. In *Dead Wave,* edited by J. Andrzejewski, pp. 79–84. Warszawa: Spółdzielnia Wydawniaza, 1947.

Iwański, Henryk. "Mówi Major Bystry" [Major Bystry speaks]. *Kultura*, Warszawa April 21, 1968, p. 6.

Kamińska, Anna. "Pisane Paznokciem" [Written with a fingernail]. *Więź*, no. 4 (April 1978): 57–85.

Kamińska-Fishman, Karyna. "Zachód Emigracyjny Rząd Oraz Polski Delegatura Wobec Sprawy Żydowskiej Podczas Wojny Światowej" [The government in exile and Delegatura, their attitudes toward Jews during the war]. *ZIH*, no. 62 (1967): 43–58.

Kann, Maria. "Martwa Fala" [Dead wave]. In *Dead Wave*, edited by J. Andrzejewski, pp. 11–19. Warszawa: Spółdzielnia Wydawnicza, 1947.

Kossak-Szczucka, Zofia. "Komu Pomagamy" [Whom do we help?]. *Prawda, Pismo Frontu Odrodzenia*, August–September 1943, pp. 7–9.

Laufer-Zorne, Fania. "Dziennik Z Czasów Okupacji Hitlerowskiej" [Occupation days diary]. *ZIH*, no. 59 (1966): 93–100.

Landau, Ludwik. "Z Dziennika" [From a diary]. *ZIH*, no. 5 (1953): 96–109.

Margules, Zantea. "Moje Życie W Tarnopolu Podczas Wojny" [My life in Tarnopol during the war]. *ZIH*, no. 36 (1960): 62–94.

Modrzewska, Krystyna. "Pamiętnik Z Okresu Okupacji" [Diary from the occupation]. *ZIH*, no. 31 (1959): 57–80.

Pełczyński, Tadeusz, and Ciołkosz, Adam. "Opór Zbrojny W Ghetcie Warszawskim W 1943." *Bellona*, London, Zeszyt 1–2, 1963, pp. 42–54.

Pfefer, Jerzy. "Moja Ucieczka Z Majdanka" [My escape from Majdanek]. *ZIH*, no. 71–72 (1969): 204–227.

Poraj, Krystyna. "Dziennik Lwowski" [Lwow diary]. *ZIH*, no. 52 (1964): 79–106.

Przyboś, Julian. "Hańba Antysemityzmu" [Disgrace of anti-Semitism]. In *Dead Wave*, edited by J. Andrzejeski, pp. 60–64. Warszawa: Spółdzieinia Wydawnicza, 1947.

Rok, Abram. "Dom Otwartych Serc" [The warmhearted home]. *Folks Sztyme*, Warszawa, June 17, 1978, pp. 10, 12.

Rok, Abram. "Genowefa I Ignacy Naruszkowie" [Genowefa and Ignacy Naruszko]. *Folks Sztyme*, Warszawa, August 5, 1978, pp. 11–12.

Rok, Abram. "Jedno Ludzkie Życie" [One human life]. *Folks Sztyme*, Warszawa, September 9, 1978, pp. 11–12.

Rok, Abram. "Telefon Łącznikowy 12-72-32" [Connecting phone]. *Folks Sztyme*, Warszawa, July 1, 1978, pp. 11–12.

Rubinstein, Stefania. "Pamiętnik Ze Lwowa" [Lwow Diary]. *ZIH*, no. 61 (1967): 87–113.

Schechter, Felicja. "Ucieczka Przed Śmiercią, Pamiętnik Z Czasów Okupacji Hitlerowskiej W Krakowie" [Escape from death during Hitler's occupation in Creaw]. *ZIH*, no. 55 (1965): 93–109.

Sendlerowa, Irena. "Ci Którzy Pomagali Żydom" Wspomienia Z Czasów Okupacji Hitlerowskiej" [Those who helped Jews during Hitler's occupation]. *ZIH*, no. 45–46 (1963): 234–247.

Sendlerowa, Irena. "O Działalnosci Kół Młodzieży Przy Komitetach Domowych W Gecie Warszawskim" [Youth Participation in the house committees in the Warsaw ghetto]. *ZIH*, no. 118 (April–June 1981): 89–118.

Sierakowski, Dawid. "Pamiętnik Z Ghetta Łódzkiego" [Diary from Lódź ghetto]. *ZIH*, no. 28 (1958): 79–110.

Switał, Stanisław. "Siedmioro Z Ulicy Promyka" [Seven from Promyko Street]. *ZIH*, no. 65–66 (1968): 207–210.

Szymańska, Zofia. "Ratunek W Klasztorze" [Rescue in the convent]. *ZIH* 4, no. 88 (1973): 33–44.

Szymańska, Zofia. "Ze Wspomień Lekarza" [From a physician's memories]. *ZIH* 4, no. 80 (1971): 51–64.

Tusk-Schweinwechslerowa, Felicja. "Cena Jednego Życia, Wspomienia Z Czasów Okupacji" [The price of a life, recalling the occupation]. *ZIH*, no. 33 (1960): 88–104.

Wizling, P. "Unserer Retter" [Our saviors]. In *Sefer Horodenko* [The book of Horodenko], edited by S. Metzger, pp. 307–321. Israel and United States: published by former residents of Horodenko and vicinity, 1963.

Wolfson, Manfred. "Zum Wiederstand gegen Hitler: Umriss eines Gruppenporträts Deutscher Retter von Juden" [Opposition to Hitler: profile of the German rescuers of Jews]. *Tradition und Neubeginn: Internationale Forschungen Deutscher Geschichte im 20, Jahrhundert* 26 (1975): 391–407.

Wyleżyńska, Aurelia. "Z Notatek Pamiętnikarskich, 1942–1943" [From diary notes]. *ZIH*, no. 45–46 (1963): 212–233.

Żabiński, Jan. *ZIH*, no. 65–66, 1968, pp. 198–199.

Zadrecki, Tadeusz. "Ludzie I . . . Ludzie—Hieny" [People and hyenas]. *Opinia* 20, no. 2 (1946): 8.

Zawiejski, Jerzy. "Nadzieja Ludzi Dobrej Woli" [Hope of people of good intentions]. In *Dead Wave*, edited by J. Andrzejewski, pp. 71–78. Warszawa: Spółdzielnia Wydawnicza, 1947.

Ziemiński, Stanisław. "Kartki Dziennika Nauczyciela W Łukowie Z Okresu Okupacji Hitlerowskiej" [Notes from a teacher's diary from Łukow during Hitler's occupation]. *ZIH*, no. 27 (1958): 105–112.

Foreign Books

Andrzejewski, Jerzy, ed. *Martwa Fala, Zbiór Artykułów O Antysemityźmie* [Dead wave]. Warszawa: Spółdzielnia Wydawnicza, 1947.

Arczyński, Marek, and Balcerak, Wieslaw. *Kryptonim "Żegota"* [Żegota]. Warszawa: Czytelnik, 1979.

Bartoszewski, Władysław. *1859 Dni Warszawy* [1859 Warsaw days]. Kraków: Wydawnictwo Znak, 1974.

Bartoszewski, Władysław. *Straceni Na Ulicach Miasta* [Perished in the streets of the city]. Warszawa: Książka I Wiedza, 1970.

Bartoszewski, Władysław, and Lewinówna, Zofia. *Ten Jest Z Ojczyzny Mojej* [This one is from my country]. Kraków: Wydawnictwo Znak, 1969.

Berenstein, Tatiana; Eisenbach, Artur; Mark, Bernard; and Rutkowski, Adam, eds. *Emmanuel Ringelblum Kronika Getta Warszawskiego, Wrzesień 1939–Styczeń 1943* [Emmanuel Ringelblum's chronicle of the Warsaw ghetto September 1939–January 1943]. Warszawa: Czytelnik, 1983.

Berenstein, Tatiana; Eisenbach, Arthur; and Rutkowski, Adam, eds. *Exterminacja Żydow Na Ziemiach Polskich W Okresie Okupacji Hitlerowskiej* [Extermination of Jews on Polish land during the German occupation], collection of documents. Warszawa: Żydowski Instytut Historyczny, 1957.

Birnbaum, Irena. *Non Omnis Moriar: Pamiętnik Z Getta Warszawskiego* [Warsaw ghetto diary]. Warszawa: Czytelnik, 1982.

Borwicz, Michael M. *Arische Papirn* [Aryan papers], 3 vols. Central Verband für Polische Idn in Argentine, Buenos Aires, 1955.

Borwicz, Michael M. *Les Vies Interdites* [Forbidden lives]. Paris: Casterman, 1969.

Borwicz, Michael M. *Organizowanie Wściekłości* [Organizing fury]. Warszawa: Wydawnictwo Ogólnpolskiej Ligi Do Walki Z Rasizmem, 1947.

Czajka, Izabella Stachowicz. *Ocalił Mnie Kowal* [I was saved by a blacksmith]. Warszawa: Czytelnik, 1956.

Czapska, Maria. *Gwiazda Dawida Dzieje Jednej Rodziny* [The star of David, the story of one family]. Londyn: Oficyna Poetów I Malarzy, 1975.

Chrześcijańskie Stowarzyszenie Społeczne [Christian charity]. Warszawa: Dzieło Miłosierdzia Chrzescijańskiego, 1961.

Choroba Głodowa [Starvation illness]. Warsaw: American Joint Distribution Committee, 1946.

Datner, Szymon. *Las Sprawiedliwych* [The forest of the righteous]. Warszawa: Książka I Wiedza, 1968.

Documenty Zbrodni I Męczeństwa [Documents about murder and suffering]. Kraków: Żydowska Komisja Historyczna W Krakowie, 1945.

Draegner, Guta. *Pemiętnik Justyny* [Justina's diary]. Kraków: Wojewódzka Żydowska Komisja Historyczna, 1946.

Grossman, Kurt *Die Unbesungenen Helden* [The unsung heroes]. Berlin: Grunewald, Arani Verlag, GmbH., 1961.

Hescheles, Janka. *Oczyma Dwunnastoletniej Dziewczyny* [With the eyes of a twelve-year-old]. Kraków: Wojewódzka Żydowska Komisja Historyczna W Krakowie, 1946.

Hirszfeld, Ludwik. *Historia Jednego Życia* [The story of one life]. Warszawa: Pax, 1957.

Hochberg-Marjanska, Maria, and Gruss, Noe, eds. *Dzieci Oskarżaja* [Children accuse]. Warszawa Centralna Żydowska Komisja Historychna W Polsce, 1947.

Hołuj, Tadeusz. *Koniec Naszego Świata* [The end of our world]. Kraków: Wydawnictwo Literackie, 1959.

Kielkowski, Roman. *Zlikwidować Na Miejscu, Z Dziejów Okupacji Hitlerowskiej W Krakowie* [To liquidate on location, Nazi occupation events in Cracaw). Kraków: Wydawnictwo Literackie, 1981.

Klukowski, Zygmunt. *Dziennik Z Lat Okupacji, Zamojszczyzny* [Wartime diary from Zamojść]. Lublin: Lubelska Spółdzielnia Wydawnicza, 1958.

Kozłowska-Balicka, Helena. *Mur Miał Dwie Strony* [The wall had two sides]. Warszawa: Wydawnictwo Ministerstwa Obrony Narodowej, 1958.

Krall, Hanna. *Zdążyć Przed Panem Bogiem* [To make it before God]. Kraków: Wydawnictwo Literackie, 1977.

Kunicki, Mikołaj. *Pamętnik Muchy* [Mucha's diary]. Warszawa: Książka I Wiedza, 1971.

Kurbacz, Jan. *List Do Wojtka* [A letter to Wojtek]. Warszawa: Ludowa Spółdzielnia Wydawnicza, 1963.

Landau, Ludwik. *Kronika Lat Wojny I Okupacji* [War chronicle], 3 vols. Warszawa: Państwowe Wydawnictwo Naukowe, 1962.

Lerski, Jerzy. *Emisariusz "Jur"* (The emissary "Jur"). Londyn: Polska Fundacja Kulturalna, 1984.

Marko-Strauss, Salomon. *Czysta Krew* [Pure blood]. Lódź: Wydawnictwo Lódzkie, 1966.

Moczarski, Kazimierz. *Rozmowy Z Katem* [Talks with a hangman]. Warszawa: Państwowy Instytut Wydawniczy, 1978.

Prekerowa, Teresa. *Konspiracyjna Rada Pomocy, Żydom W Warszawie 1942–1945*

[Illegal help to Jews in Warsaw]. Warszawa: Państwowy Instytut Wydawniczy, 1982.

Rubinowicz, Dawid. *Pamietnik Dawida Rubinowicza* [David's diary]. Warszawa: Książka I Wiedza, 1960.

Sakowska, Ruta, ed. *Archiwum Ringelbluma Ghetto Warszawskie, Lipiec 1942– Styczeń 1943* [Ringelblum's archives of the Warsaw ghetto, July 1942–January 1943]. Warszawa: Państwowe Wydawnictwo, 1980.

Sierakowiak, Dawid. *Dziennik Dawida Sierakowiaka* [David Sierakowiak's diary]. Warszawa: Iskry, 1960.

Smolski, Władysław. *Za To Graziła Śmierć* [Death threatened for this]. Warszawa: Pax, 1981.

Smolski, Władysław. *Zaklęte Lata* [Cursed years]. Warszawa: Instytut Wydwaniczy Pax, 1964.

Sobiesiak, Józef. *Ziemia Płonie* [The soil burns]. Lublin: Wydawnictwo Lubelskie, 1974.

Szarota, Okupowanej Warszawy Dzień Powszedni, Studium Historyczne [A common day in occupied Warsaw]. Warszawa: Czytelnik, 1973.

Szpilman, Władysław. *Śmierć Miasta Pamiętniki Władysława Szpilmana* [Death of a city]. Warszawa: Spółdzielnia Wydawnicza, 1946.

Szyfman, Arnold. *Moja Tułaczka Wojenna* [My wartime wanderings]. Warszawa: Wydawnictwo Obrony Narodowej, 1946.

Szymańska, *Byłam Tylko Lekarzem* [I was only a physician]. Warszawa: Pax, 1979.

Wanat, Leon. *Za Murami Pawiaka* [Behind Pawiak's walls]. Warszawa: Książka I Wiedza, 1958.

Wroński, Stanisław, and Zwolakowa, Maria. *Polacy I Żydzi, 1939–1945* [Poles and Jews]. Warszawa: Książka I Wiedza, 1971.

Zabęcki, Franiszek. *Wspomnienia Dawne I Nowe* [Old and new memories]. Warszawa: Pax, 1977.

Zagórski, Wacław. *Wolność W Niewoli* [Freedom in slavery]. London: Nakładem Autora I Czytelników Przedstawiciela, 1971.

Unpublished Sources

1. Jewish Historical Institute, Warsaw
    A. Collections of eyewitness testimonies on the Holocaust
    B. Collections of unpublished wartime memoirs
    C. Collection of documents about the Council for Aid to Jews (Żegota)
2. Yad Vashem, Martyrs' and Heroes Memorial Authority in Jerusalem
    A. Collections of Holocaust eyewitness testimonies
    B. Collections of unpublished wartime memoirs
    C. Requests for Yad Vashem distinctions
3. YIVO Institute for Jewish Research, New York
    A. Collections of eyewitness testimonies on the Holocaust, Series I, II, and III
    B. Boder, David P. *Topical Autobiographies of Displaced Persons*, 16 vols., 1956 (mimeographed)
    C. Territorial Collection on Poland, Vilno Archives Record Group, no. 28, 1931.

Index